PRE-ALGEBRA TUTOR

Everything You Need to Help Achieve

an Excellent Score

By

Reza Nazari

&

Ava Ross

EffortlessMath

About Effortless Math Education

Effortless Math Education operates the www.effortlessmath.com website, which prepares and publishes Test prep and Mathematics learning resources. Effortless Math authors' team strives to prepare and publish the best quality Mathematics learning resources to make learning Math easier for all. We Help Students Learn to Love Mathematics.

All inquiries should be addressed to:
info@effortlessMath.com
www.EffortlessMath.com

ISBN: 978-1-64612-865-5

Published by: **Effortless Math Education**

for Online Math Practice Visit www.EffortlessMath.com

Pre-Algebra TUTOR

All the Tools You Need to Succeed on the Pre-Algebra test 2020!

Feeling anxious about the Pre-Algebra? Not sure your math skills are up to the task? Don't worry, *Pre-Algebra Tutor* has you covered!

Focusing on proven test-taking strategies, easy-to-understand math principles, and professional guidance, *Pre-Algebra Tutor* is your comprehensive study guide for the Pre-Algebra test!

Each chapter includes a study-guide formatted review and quizzes to check your comprehension on the topics covered. With this self-study guide, it's like having your own tutor for a fraction of the cost!

What does the Pre-Algebra Tutor offer?

- Content 100% aligned with the 2020 Pre-Algebra test

- **Step-by-Step guides** to all Pre-Algebra concepts and topics covered in the 2020 test

- **Over 500 additional Pre-Algebra practice questions** featuring multiple-choice and grid-in formats with answers grouped by topic, so you can focus on your weak areas

- Abundant Math skill-building exercises to help test-takers approach different question types that might be unfamiliar to them

The surest way to succeed on the Pre-Algebra Test is with intensive practice in every math topic tested—and that's what exactly what you'll get! With the *Pre-Algebra Tutor*, you'll have everything you need to ace the Pre-Algebra right in your hands. ***Start studying today!***

About the Author

Reza Nazari is the author of more than 100 Math learning books including:

- ❖ **Math and Critical Thinking Challenges:** For the Middle and High School Student
- ❖ **ACT Math in 30 Days**
- ❖ **ASVAB Math Workbook 2018 - 2019**
- ❖ **Effortless Math Education Workbooks**
- ❖ **and many more Mathematics books**

Reza is also an experienced Math instructor and a test–prep expert who has been tutoring students since 2008. Reza is the founder of Effortless Math Education, a tutoring company that has helped many students raise their standardized test scores—and attend the colleges of their dreams. Reza provides an individualized custom learning plan and the personalized attention that makes a difference in how students view math.

You can contact Reza via email at:
reza@EffortlessMath.com

Find Reza's professional profile at:
goo.gl/zoC9rJ

Contents

11 GEOMETRY AND SOLID FIGURES   109

12 STATISTICS   121

CHAPTER 1:

FRACTIONS AND MIXED NUMBERS

Math Topics that you'll learn in this chapter:

▶ Simplifying Fractions

▶ Adding and Subtracting Fractions

▶ Multiplying and Dividing Fractions

▶ Adding Mixed Numbers

▶ Subtracting Mixed Numbers

▶ Multiplying Mixed Numbers

▶ Dividing Mixed Numbers

SIMPLIFYING FRACTIONS

☑ A fraction contains two numbers separated by a bar between them. The bottom number, called the denominator, is the total number of equally divided portions in one whole. The top number, called the numerator, is how many portions you have. And the bar represents the operation of division.

☑ Simplifying a fraction means reducing it to the lowest terms. To simplify a fraction, evenly divide both the top and bottom of the fraction by $2, 3, 5, 7$, etc.

☑ Continue until you can't go any further.

Examples:

Example 1. Simplify $\frac{12}{30}$

Solution: To simplify $\frac{12}{30}$, find a number that both 12 and 30 are divisible by. Both are divisible by 6. Then: $\frac{12}{30} = \frac{12 \div 6}{30 \div 6} = \frac{2}{5}$

Example 2. Simplify $\frac{64}{80}$

Solution: To simplify $\frac{64}{80}$, find a number that both 64 and 80 are divisible by. Both are divisible by 8 and 16. Then: $\frac{64}{80} = \frac{64 \div 8}{80 \div 8} = \frac{8}{10}$, 8 and 10 are divisible by 2, then: $\frac{8}{10} = \frac{4}{5}$ or $\frac{64}{80} = \frac{64 \div 16}{80 \div 16} = \frac{4}{5}$

Example 3. Simplify $\frac{20}{60}$

Solution: To simplify $\frac{20}{60}$, find a number that both 20 and 60 are divisible by. Both are divisible by 20, then: $\frac{20}{60} = \frac{20 \div 20}{60 \div 20} = \frac{1}{3}$

ADDING AND SUBTRACTING FRACTIONS

☑ For "like" fractions (fractions with the same denominator), add or subtract the numerators (top numbers) and write the answer over the common denominator (bottom numbers).

☑ Adding and Subtracting fractions with the same denominator:

$$\frac{a}{b} + \frac{c}{b} = \frac{a+c}{b} \qquad \frac{a}{b} - \frac{c}{b} = \frac{a-c}{b}$$

☑ Find equivalent fractions with the same denominator before you can add or subtract fractions with different denominators.

☑ Adding and Subtracting fractions with different denominators:

$$\frac{a}{b} + \frac{c}{d} = \frac{ad+bc}{bd} \qquad \frac{a}{b} - \frac{c}{d} = \frac{ad-bc}{bd}$$

Examples:

Example 1. Find the sum. $\frac{3}{4} + \frac{1}{3} =$

Solution: These two fractions are "unlike" fractions. (they have different denominators). Use this formula: $\frac{a}{b} + \frac{c}{d} = \frac{ad+cb}{bd}$

Then: $\frac{3}{4} + \frac{1}{3} = \frac{(3)(3)+(4)(1)}{4 \times 3} = \frac{9+4}{12} = \frac{13}{12}$

Example 2. Find the difference. $\frac{4}{5} - \frac{3}{7} =$

Solution: For "unlike" fractions, find equivalent fractions with the same denominator before you can add or subtract fractions with different denominators. Use this formula:

$\frac{a}{b} - \frac{c}{d} = \frac{ad-bc}{bd}$

$\frac{4}{5} - \frac{3}{7} = \frac{(4)(7)-(3)(5)}{5 \times 7} = \frac{28-15}{35} = \frac{13}{35}$

MULTIPLYING AND DIVIDING FRACTIONS

☑ Multiplying fractions: multiply the top numbers and multiply the bottom numbers. Simplify if necessary. $\dfrac{a}{b} \times \dfrac{c}{d} = \dfrac{a \times c}{b \times d}$

☑ Dividing fractions: Keep, Change, Flip

☑ Keep the first fraction, change the division sign to multiplication, and flip the numerator and denominator of the second fraction. Then, solve!

$$\frac{a}{b} \div \frac{c}{d} = \frac{a}{b} \times \frac{d}{c} = \frac{a \times d}{b \times c}$$

Examples:

Example 1. Multiply. $\dfrac{5}{8} \times \dfrac{2}{3} =$

Solution: Multiply the top numbers and multiply the bottom numbers. $\dfrac{5}{8} \times \dfrac{2}{3} = \dfrac{5 \times 2}{8 \times 3} = \dfrac{10}{24}$, simplify: $\dfrac{10}{24} = \dfrac{10 \div 2}{24 \div 2} = \dfrac{5}{12}$

Example 2. Solve. $\dfrac{1}{3} \div \dfrac{4}{7} =$

Solution: Keep the first fraction, change the division sign to multiplication, and flip the numerator and denominator of the second fraction.
Then: $\dfrac{1}{3} \div \dfrac{4}{7} = \dfrac{1}{3} \times \dfrac{7}{4} = \dfrac{1 \times 7}{3 \times 4} = \dfrac{7}{12}$

Example 3. Calculate. $\dfrac{3}{5} \times \dfrac{2}{3} =$

Solution: Multiply the top numbers and multiply the bottom numbers. $\dfrac{3}{5} \times \dfrac{2}{3} = \dfrac{3 \times 2}{5 \times 3} = \dfrac{6}{15}$, simplify: $\dfrac{6}{15} = \dfrac{6 \div 3}{15 \div 3} = \dfrac{2}{5}$

Example 4. Solve. $\dfrac{1}{4} \div \dfrac{5}{6} =$

Solution: Keep the first fraction, change the division sign to multiplication, and flip the numerator and denominator of the second fraction.
Then: $\dfrac{1}{4} \div \dfrac{5}{6} = \dfrac{1}{4} \times \dfrac{6}{5} = \dfrac{1 \times 6}{4 \times 5} = \dfrac{6}{20}$, simplify: $\dfrac{6}{20} = \dfrac{6 \div 2}{20 \div 2} = \dfrac{3}{10}$

ADDING MIXED NUMBERS

Use the following steps for adding mixed numbers:

☑ Add whole numbers of the mixed numbers.

☑ Add the fractions of the mixed numbers.

☑ Find the Least Common Denominator (LCD) if necessary.

☑ Add whole numbers and fractions.

☑ Write your answer in lowest terms.

Examples:

Example 1. Add mixed numbers. $3\frac{1}{3} + 1\frac{4}{5} =$

Solution: Let's rewriting our equation with parts separated, $3\frac{1}{3} + 1\frac{4}{5} = 3 + \frac{1}{3} + 1 + \frac{4}{5}$. Now, add whole number parts: $3 + 1 = 4$

Add the fraction parts $\frac{1}{3} + \frac{4}{5}$. Rewrite to solve with the equivalent fractions.

$\frac{1}{3} + \frac{4}{5} = \frac{5}{15} + \frac{12}{15} = \frac{17}{15}$. The answer is an improper fraction (numerator is bigger

than denominator). Convert the improper fraction into a mixed number:

$\frac{17}{15} = 1\frac{2}{15}$. Now, combine the whole and fraction parts: $4 + 1\frac{2}{15} = 5\frac{2}{15}$

Example 2. Find the sum. $1\frac{2}{5} + 2\frac{1}{2} =$

Solution: Rewriting our equation with parts separated, $1 + \frac{2}{5} + 2 + \frac{1}{2}$. Add

the whole number parts:

$1 + 2 = 3$. Add the fraction parts: $\frac{2}{5} + \frac{1}{2} = \frac{4}{10} + \frac{5}{10} = \frac{9}{10}$

Now, combine the whole and fraction parts: $3 + \frac{9}{10} = 3\frac{9}{10}$

Subtract Mixed Numbers

Use these steps for subtracting mixed numbers.

☑ Convert mixed numbers into improper fractions. $a\frac{c}{b} = \frac{ab+c}{b}$

☑ Find equivalent fractions with the same denominator for unlike fractions. (fractions with different denominators)

☑ Subtract the second fraction from the first one. $\frac{a}{b} - \frac{c}{d} = \frac{ad-bc}{bd}$

☑ Write your answer in lowest terms.

☑ If the answer is an improper fraction, convert it into a mixed number.

Examples:

Example 1. Subtract. $3\frac{4}{5} - 1\frac{3}{4} =$

Solution: Convert mixed numbers into fractions: $3\frac{4}{5} = \frac{3\times5+4}{5} = \frac{19}{5}$ and $1\frac{3}{4} = \frac{1\times4+3}{4} = \frac{7}{4}$

These two fractions are "unlike" fractions. (they have different denominators). Find equivalent fractions with the same denominator. Use this formula: $\frac{a}{b} - \frac{c}{d} = \frac{ad-bc}{bd}$

$\frac{19}{5} - \frac{7}{4} = \frac{(19)(4)-(5)(7)}{5\times4} = \frac{76-35}{20} = \frac{41}{20}$, the answer is an improper fraction, convert it into a mixed number. $\frac{41}{20} = 2\frac{1}{20}$

Example 2. Subtract. $4\frac{3}{8} - 1\frac{1}{2} =$

Solution: Convert mixed numbers into fractions: $4\frac{3}{8} = \frac{4\times8+3}{8} = \frac{35}{8}$ and $1\frac{1}{2} = \frac{1\times2+1}{4} = \frac{3}{2}$

Find equivalent fractions: $\frac{3}{2} = \frac{12}{8}$. Then: $4\frac{3}{8} - 1\frac{1}{2} = \frac{35}{8} - \frac{12}{8} = \frac{23}{8}$

The answer is an improper fraction, convert it into a mixed number.

$$\frac{23}{8} = 2\frac{7}{8}$$

MULTIPLYING MIXED NUMBERS

Use the following steps for multiplying mixed numbers:

☑ Convert the mixed numbers into fractions. $a\frac{c}{b} = a + \frac{c}{b} = \frac{ab+c}{b}$

☑ Multiply fractions. $\frac{a}{b} \times \frac{c}{d} = \frac{a \times c}{b \times d}$

☑ Write your answer in lowest terms.

☑ If the answer is an improper fraction (numerator is bigger than denominator), convert it into a mixed number.

Examples:

Example 1. Multiply. $3\frac{1}{3} \times 4\frac{1}{6} =$

Solution: Convert mixed numbers into fractions, $3\frac{1}{3} = \frac{3 \times 3 + 1}{3} = \frac{10}{3}$ and $4\frac{1}{6} = \frac{4 \times 6 + 1}{6} = \frac{25}{6}$

Apply the fractions rule for multiplication, $\frac{10}{3} \times \frac{25}{6} = \frac{10 \times 25}{3 \times 6} = \frac{250}{18}$

The answer is an improper fraction. Convert it into a mixed number. $\frac{250}{18} = 13\frac{8}{9}$

Example 2. Multiply. $2\frac{1}{2} \times 3\frac{2}{3} =$

Solution: Converting mixed numbers into fractions, $2\frac{1}{2} \times 3\frac{2}{3} = \frac{5}{2} \times \frac{11}{3}$

Apply the fractions rule for multiplication, $\frac{5}{2} \times \frac{11}{3} = \frac{5 \times 11}{2 \times 3} = \frac{55}{6} = 9\frac{1}{6}$

Example 3. Multiply mixed numbers. $2\frac{1}{3} \times 2\frac{1}{2} =$

Solution: Converting mixed numbers to fractions, $2\frac{1}{3} = \frac{7}{3}$ and $2\frac{1}{2} = \frac{5}{2}$. Multiply two fractions:

$$\frac{7}{3} \times \frac{5}{2} = \frac{7 \times 5}{3 \times 2} = \frac{35}{6} = 5\frac{5}{6}$$

DIVIDING MIXED NUMBERS

Use the following steps for dividing mixed numbers:

☑ Convert the mixed numbers into fractions. $a\frac{c}{b} = a + \frac{c}{b} = \frac{ab+c}{b}$

☑ Divide fractions: Keep, Change, Flip: Keep the first fraction, change the division sign to multiplication, and flip the numerator and denominator of the second fraction. Then, solve! $\frac{a}{b} \div \frac{c}{d} = \frac{a}{b} \times \frac{d}{c} = \frac{a \times d}{b \times c}$

☑ Write your answer in lowest terms.

☑ If the answer is an improper fraction (numerator is bigger than denominator), convert it into a mixed number.

Examples:

Example 1. Solve. $3\frac{2}{3} \div 2\frac{1}{2}$

Solution: Convert mixed numbers into fractions: $3\frac{2}{3} = \frac{3 \times 3 + 2}{3} = \frac{11}{3}$ and $2\frac{1}{2} = \frac{2 \times 2 + 1}{2} = \frac{5}{2}$

Keep, Change, Flip: $\frac{11}{3} \div \frac{5}{2} = \frac{11}{3} \times \frac{2}{5} = \frac{11 \times 2}{3 \times 5} = \frac{22}{15}$. The answer is an improper fraction. Convert it into a mixed number: $\frac{22}{15} = 1\frac{7}{15}$

Example 2. Solve. $3\frac{4}{5} \div 1\frac{5}{6}$

Solution: Convert mixed numbers to fractions, then solve:

$3\frac{4}{5} \div 1\frac{5}{6} = \frac{19}{5} \div \frac{11}{6} = \frac{19}{5} \times \frac{6}{11} = \frac{114}{55} = 2\frac{4}{55}$

Example 3. Solve. $2\frac{2}{7} \div 2\frac{3}{5}$

Solution: Converting mixed numbers to fractions: $3\frac{4}{5} \div 1\frac{5}{6} = \frac{16}{7} \div \frac{13}{5}$

Keep, Change, Flip: $\frac{16}{7} \div \frac{13}{5} = \frac{16}{7} \times \frac{5}{13} = \frac{16 \times 5}{7 \times 13} = \frac{80}{91}$

CHAPTER 1: PRACTICES

✎ Simplify each fraction.

1) $\frac{16}{24} =$

2) $\frac{28}{70} =$

3) $\frac{30}{105} =$

4) $\frac{40}{35} =$

5) $\frac{48}{56} =$

6) $\frac{6}{120} =$

7) $\frac{15}{100} =$

8) $\frac{45}{54} =$

✎ Find the sum or difference.

9) $\frac{4}{12} + \frac{3}{12} =$

10) $\frac{5}{4} + \frac{1}{12} =$

11) $\frac{3}{6} + \frac{2}{5} =$

12) $\frac{8}{25} - \frac{3}{25} =$

13) $\frac{5}{3} - \frac{2}{9} =$

14) $\frac{3}{2} - \frac{3}{4} =$

15) $\frac{4}{3} - \frac{6}{5} =$

16) $\frac{5}{12} + \frac{3}{5} =$

✎ Find the products or quotients.

17) $\frac{9}{5} \div \frac{3}{2} =$

18) $\frac{8}{7} \div \frac{4}{3} =$

19) $\frac{6}{4} \times \frac{8}{5} =$

20) $\frac{7}{2} \times \frac{4}{9} =$

✎ Find the sum.

21) $2\frac{1}{3} + 1\frac{4}{5} =$

22) $4\frac{3}{7} + 3\frac{3}{4} =$

23) $2\frac{3}{4} + 3\frac{1}{3} =$

24) $1\frac{1}{4} + 3\frac{1}{2} =$

25) $2\frac{5}{7} + 2\frac{1}{3} =$

26) $4\frac{2}{9} + 2\frac{1}{2} =$

✍ Find the difference.

27) $4\frac{2}{9} - 3\frac{1}{7} =$

28) $3\frac{3}{4} - 2\frac{1}{8} =$

29) $3\frac{2}{7} - 2\frac{4}{9} =$

30) $8\frac{3}{4} - 2\frac{1}{8} =$

31) $5\frac{5}{6} - 3\frac{1}{24} =$

32) $7\frac{3}{10} - 4\frac{4}{5} =$

33) $8\frac{1}{6} - 3\frac{2}{3} =$

34) $14\frac{9}{10} - 8\frac{4}{5} =$

✍ Find the products.

35) $2\frac{1}{9} \times 2\frac{5}{6} =$

36) $2\frac{3}{4} \times 4\frac{1}{9} =$

37) $1\frac{2}{7} \times 1\frac{5}{6} =$

38) $3\frac{2}{9} \times 1\frac{6}{5} =$

39) $3\frac{2}{3} \times 2\frac{3}{5} =$

40) $2\frac{5}{6} \times 3\frac{1}{9} =$

41) $3\frac{4}{5} \times 1\frac{1}{6} =$

42) $4\frac{1}{5} \times 1\frac{2}{7} =$

✍ Solve.

43) $8\frac{3}{4} \div 4\frac{1}{3} =$

44) $4\frac{2}{5} \div 1\frac{2}{9} =$

45) $6\frac{1}{2} \div 2\frac{1}{3} =$

46) $7\frac{1}{6} \div 3\frac{4}{9} =$

47) $2\frac{1}{4} \div 1\frac{1}{8} =$

48) $3\frac{2}{5} \div 1\frac{1}{10} =$

49) $4\frac{1}{2} \div 2\frac{2}{3} =$

50) $11\frac{1}{3} \div 2\frac{2}{9} =$

CHAPTER 1: ANSWERS

1) $\frac{2}{3}$

2) $\frac{2}{5}$

3) $\frac{2}{7}$

4) $\frac{8}{7}$

5) $\frac{6}{7}$

6) $\frac{1}{20}$

7) $\frac{3}{20}$

8) $\frac{5}{6}$

9) $\frac{7}{12}$

10) $\frac{4}{3}$

11) $\frac{9}{10}$

12) $\frac{1}{5}$

13) $\frac{13}{9} = 1\frac{4}{9}$

14) $\frac{3}{4}$

15) $\frac{2}{15}$

16) $\frac{61}{60} = 1\frac{1}{60}$

17) $\frac{6}{5}$

18) $\frac{6}{7}$

19) $\frac{12}{5} = 2\frac{2}{5}$

20) $\frac{14}{9} = 1\frac{5}{9}$

21) $4\frac{2}{15}$

22) $8\frac{5}{28}$

23) $6\frac{1}{12}$

24) $4\frac{3}{4}$

25) $5\frac{1}{21}$

26) $6\frac{13}{18}$

27) $1\frac{5}{63}$

28) $1\frac{5}{8}$

29) $\frac{53}{63}$

30) $6\frac{5}{8}$

31) $2\frac{19}{24}$

32) $2\frac{1}{2}$

33) $4\frac{1}{2}$

34) $6\frac{1}{10}$

35) $5\frac{53}{54}$

36) $11\frac{11}{63}$

37) $2\frac{5}{14}$

38) $7\frac{4}{45}$

39) $9\frac{18}{15}$

40) $8\frac{22}{27}$

41) $4\frac{13}{30}$

42) $5\frac{2}{5}$

43) $2\frac{1}{52}$

44) $3\frac{3}{5}$

45) $2\frac{11}{14}$

46) $2\frac{5}{62}$

47) 2

48) $3\frac{1}{11}$

49) $1\frac{11}{16}$

50) $5\frac{1}{10}$

CHAPTER 2:

DECIMALS

Math Topics that you'll learn in this chapter:

▶ Comparing Decimals

▶ Rounding Decimals

▶ Adding and Subtracting Decimals

▶ Multiplying and Dividing Decimals

COMPARING DECIMALS

☑ A decimal is a fraction written in a special form. For example, instead of writing $\frac{1}{2}$ you can write 0.5

☑ A Decimal Number contains a Decimal Point. It separates the whole number part from the fractional part of a decimal number.

☑ Let's review decimal place values: Example: **53.9861**

5: tens   3: ones   9: tenths

8: hundredths   6: thousandths   1: tens thousandths

☑ To compare decimals, compare each digit of two decimals in the same place value. Start from left. Compare hundreds, tens, ones, tenth, hundredth, etc.

☑ To compare numbers, use these symbols:

Equal to =,   Less than <,   Greater than >

Greater than or equal ≥,   Less than or equal ≤

Examples:

Example 1. Compare 0.60 and 0.06.

Solution: 0.60 *is greater than* 0.06, because the tenth place of 0.60 is 6, but the tenth place of 0.06 is zero. Then: 0.60 > 0.06

Example 2. Compare 0.0815 and 0.815.

Solution: 0.815 *is greater than* 0.0815, because the tenth place of 0.815 is 8, but the tenth place of 0.0815 is zero. Then: 0.0815 < 0.815

ROUNDING DECIMALS

✅ We can round decimals to a certain accuracy or number of decimal places. This is used to make calculations easier to do and results easier to understand when exact values are not too important.

✅ First, you'll need to remember your place values: For example: **12.4869**

1: tens	2: ones	4: tenths
8: hundredths	6: thousandths	9: tens thousandths

✅ To round a decimal, first find the place value you'll round to.

✅ Find the digit to the right of the place value you're rounding to. If it is 5 or bigger, add 1 to the place value you're rounding to and remove all digits on its right side. If the digit to the right of the place value is less than 5, keep the place value and remove all digits on the right.

Examples:

Example 1. Round 1.9278 to the thousandth place value.

Solution: First, look at the next place value to the right, (tens thousandths). It's 8 and it is greater than 5. Thus add 1 to the digit in the thousandth place. The thousandth place is 7. $\rightarrow 7 + 1 = 8$, then, The answer is 1.928

Example 2. Round 9.4126 to the nearest hundredth.

Solution: First, look at the digit to the right of hundredth (thousandths place value). It's 2 and it is less than 5, thus remove all the digits to the right of hundredth place. Then, the answer is 9.41

ADDING AND SUBTRACTING DECIMALS

☑ Line up the decimal numbers.

☑ Add zeros to have the same number of digits for both numbers if necessary.

☑ Remember your place values: For example: 73.5196

    7: tens   3: ones   5: tenths

    1: hundredths   9: thousandths   6: tens thousandths

☑ Add or subtract using column addition or subtraction.

Examples:

Example 1. Add. $1.8 + 3.12$

Solution: First, line up the numbers: $\begin{array}{r} 1.8 \\ + 3.12 \\ \hline \end{array} \to$ Add a zero to have the same number of digits for both numbers. $\begin{array}{r} 1.80 \\ + 3.12 \\ \hline \end{array} \to$ Start with the hundredths place: $0 + 2 = 2$, $\begin{array}{r} 1.80 \\ + 3.12 \\ \hline 2 \end{array} \to$ Continue with tenths place: $8 + 1 = 9$, $\begin{array}{r} 1.80 \\ + 3.12 \\ \hline .92 \end{array} \to$ Add the ones place: $3 + 1 = 4$, $\begin{array}{r} 1.80 \\ + 3.12 \\ \hline 4.92 \end{array}$

Example 2. Find the difference. $3.67 - 2.23$

Solution: First, line up the numbers: $\begin{array}{r} 3.67 \\ - 2.23 \\ \hline \end{array} \to$ Start with the hundredths place: $7 - 3 = 4$, $\begin{array}{r} 3.67 \\ - 2.23 \\ \hline 4 \end{array} \to$ Continue with tenths place. $6 - 2 = 4$, $\begin{array}{r} 3.67 \\ - 2.23 \\ \hline .44 \end{array} \to$ Subtract the ones place. $3 - 2 = 1$, $\begin{array}{r} 3.67 \\ - 2.23 \\ \hline 1.44 \end{array}$

MULTIPLYING AND DIVIDING DECIMALS

For multiplying decimals:

☑ Ignore the decimal point and set up and multiply the numbers as you do with whole numbers.

☑ Count the total number of decimal places in both of the factors.

☑ Place the decimal point in the product.

For dividing decimals:

☑ If the divisor is not a whole number, move the decimal point to the right to make it a whole number. Do the same for the dividend.

☑ Divide similar to whole numbers.

Examples:

Example 1. Find the product. $0.81 \times 0.32 =$

Solution: Set up and multiply the numbers as you do with whole numbers. Line up the numbers: $\begin{array}{r} 81 \\ \times 32 \\ \hline \end{array}$ → Start with the ones place then continue with other digits → $\begin{array}{r} 81 \\ \times 32 \\ \hline 2,592 \end{array}$. Count the total number of decimal places in both of the factors. There are four decimals digits. (two for each factor 0.81 and 0.32) Then: $0.81 \times 0.32 = 0.2592$

Example 2. Find the quotient. $1.60 \div 0.4 =$

Solution: The divisor is not a whole number. Multiply it by 10 to get 4: → $0.4 \times 10 = 4$

Do the same for the dividend to get 16. → $1.60 \times 10 = 1.6$

Now, divide $16 \div 4 = 4$. The answer is 4.

CHAPTER 2: PRACTICES

✎ Compare. Use >, =, and <

1) 0.55 ☐ 0.055

2) 0.34 ☐ 0.33

3) 0.66 ☐ 0.59

4) 2.650 ☐ 2.65

5) 2.34 ☐ 2.67

6) 2.46 ☐ 2.05

7) 0.16 ☐ 0.025

8) 5.05 ☐ 50.5

9) 1.020 ☐ 1.02

10) 3.022 ☐ 3.3

11) 1.400 ☐ 1.60

12) 3.44 ☐ 4.3

13) 0.380 ☐ 3.03

14) 2.081 ☐ 2.63

✎ Round each decimal to the nearest whole number.

15) 10.57

16) 4.8

17) 29.7

18) 32.58

19) 7.5

20) 8.87

21) 56.23

22) 6.39

23) 18.63

24) 25.56

25) 28.49

26) 12.67

27) 49.9

28) 17.77

29) 3.44

30) 55.56

✎ Find the sum or difference.

31) $25.31 + 56.37 =$

32) $78.32 - 65.10 =$

33) $65.80 + 14.26 =$

34) $90.24 - 53.81 =$

35) $76.41 - 49.27 =$

36) $45.39 + 17.86 =$

37) $56.02 + 30.60 =$

38) $67.01 - 28.40 =$

39) $75.14 - 25.96 =$

40) $37.52 + 13.50 =$

41) $84.71 - 54.18 =$

42) $24.12 + 29.84 =$

43) $50.59 - 46.25 =$

44) $63.13 + 21.14 =$

45) $45.23 - 35.17 =$

46) $18.02 + 30.40 =$

✎ Find the product or quotient.

47) $1.4 \times 3.2 =$

48) $8.2 \div 0.2 =$

49) 4.12×3.5

50) $6.8 \div 1.7 =$

51) $5.8 \times 0.5 =$

52) $1.54 \div 0.5 =$

53) $1.4 \times 3.2 =$

54) $5.8 \div 0.2 =$

55) $6.4 \times 7.3 =$

56) $0.3 \times 3.2 =$

57) $7.5 \times 5.6 =$

58) $45.6 \div 0.8 =$

59) $1.9 \times 5.8 =$

60) $6.74 \times 2.5 =$

61) $56.08 \div 0.2 =$

62) $36.2 \times 3.6 =$

CHAPTER 2: ANSWERS

1) >	22) 6	43) 4.34
2) >	23) 19	44) 84.27
3) >	24) 26	45) 10.06
4) =	25) 28	46) 48.42
5) <	26) 13	47) 4.48
6) >	27) 50	48) 41
7) >	28) 18	49) 14.42
8) <	29) 3	50) 4
9) =	30) 56	51) 2.9
10) <	31) 81.68	52) 3.08
11) <	32) 13.22	53) 4.48
12) <	33) 80.06	54) 29
13) <	34) 36.43	55) 46.72
14) <	35) 27.14	56) 0.96
15) 11	36) 63.25	57) 42
16) 5	37) 86.62	58) 57
17) 30	38) 38.61	59) 11.02
18) 33	39) 49.18	60) 16.85
19) 8	40) 51.02	61) 280.4
20) 9	41) 30.53	62) 130.32
21) 56	42) 53.96	

CHAPTER 3:

INTEGERS AND ORDER OF OPERATIONS

Math Topics that you'll learn in this chapter:

▶ Adding and Subtracting Integers

▶ Multiplying and Dividing Integers

▶ Order of Operations

▶ Integers and Absolute Value

ADDING AND SUBTRACTING INTEGERS

☑ Integers include zero, counting numbers, and the negative of the counting numbers. $\{\dots, -3, -2, -1, 0, 1, 2, 3, \dots\}$

☑ Add a positive integer by moving to the right on the number line. (you will get a bigger number)

☑ Add a negative integer by moving to the left on the number line. (you will get a smaller number)

☑ Subtract an integer by adding its opposite.

Examples:

Example 1. Solve. $(-4) - 5 =$

Solution: Keep the first number and convert the sign of the second number to its opposite. (change subtraction into addition. Then: $(-4) + 5 = 1$

Example 2. Solve. $11 + (8 - 19) =$

Solution: First, subtract the numbers in brackets, $8 - 19 = -11$.
Then: $11 + (-11) = \rightarrow$ change addition into subtraction: $11 - 11 = 0$

Example 3. Solve. $5 - 14 - 3 =$

Solution: First, subtract the numbers in brackets, $-14 - 3 = -17$
Then: $5 - 17 = \rightarrow$ change subtraction into addition: $5 + 17 = 22$

Example 4. Solve. $10 + (-6 - 15) =$

Solution: First, subtract the numbers in brackets, $-6 - 15 = -21$
Then: $10 + (-21) = \rightarrow$ change addition into subtraction: $10 - 21 = -11$

MULTIPLYING AND DIVIDING INTEGERS

Use the following rules for multiplying and dividing integers:

☑ (negative) × (negative) = positive

☑ (negative) ÷ (negative) = positive

☑ (negative) × (positive) = negative

☑ (negative) ÷ (positive) = negative

☑ (positive) × (positive) = positive

☑ (positive) ÷ (negative) = negative

Examples:

Example 1. Solve. $2 \times (-3) =$

Solution: Use this rule: (positive) × (negative) = negative.
Then: $(2) \times (-3) = -6$

Example 2. Solve. $(-5) + (-27 \div 9) =$

Solution: First, divide -27 by 9, the numbers in brackets, use this rule:
(negative) ÷ (positive) = negative. Then: $-27 \div 9 = -3$
$(-5) + (-27 \div 9) = (-5) + (-3) = -5 - 3 = -8$

Example 3. Solve. $(15 - 17) \times (-8) =$

Solution: First, subtract the numbers in brackets,
$15 - 17 = -2 \rightarrow (-2) \times (-8) =$
Now use this rule: (negative) × (negative) = positive $\rightarrow (-2) \times (-8) = 16$

Example 4. Solve. $(16 - 10) \div (-2) =$

Solution: First, subtract the numbers in brackets,
$16 - 10 = 6 \rightarrow (6) \div (-2) =$
Now use this rule: (positive) ÷ (negative) = negative $\rightarrow (6) \div (-2) = -3$

ORDER OF OPERATIONS

☑ In Mathematics, "operations" are addition, subtraction, multiplication, division, exponentiation (written as b^n), and grouping;

☑ When there is more than one math operation in an expression, use PEMDAS: (to memorize this rule, remember the phrase "Please Excuse My Dear Aunt Sally".)

❖ Parentheses

❖ Exponents

❖ Multiplication and Division (from left to right)

❖ Addition and Subtraction (from left to right)

Examples:

Example 1. Calculate. $(3 + 5) \div (3^2 \div 9) =$

Solution: First, simplify inside parentheses: $(8) \div (9 \div 9) = (8) \div (1)$, Then: $(8) \div (1) = 8$

Example 2. Solve. $(7 \times 8) - (12 - 4) =$

Solution: First, calculate within parentheses: $(7 \times 8) - (12 - 4) = (56) - (8)$, Then: $(56) - (8) = 48$

Example 3. Calculate. $-2[(8 \times 9) \div (2^2 \times 2)] =$

Solution: First, calculate within parentheses: $-2[(72) \div (4 \times 2)] = -2[(72) \div (8)] = -2[9]$ multiply -2 and 9. Then: $-2[9] = -18$

Example 4. Solve. $(14 \div 7) + (-13 + 8) =$

Solution: First, calculate within parentheses: $(14 \div 7) + (-13 + 8) = (2) + (-5)$ Then: $(2) - (5) = -3$

INTEGERS AND ABSOLUTE VALUE

☑ The absolute value of a number is its distance from zero, in either direction, on the number line. For example, the distance of 9 and −9 from zero on number line is 9.

☑ The absolute value of an integer is the numerical value without its sign. (negative or positive)

☑ The vertical bar is used for absolute value as in $|x|$.

☑ The absolute value of a number is never negative; because it only shows, "how far the number is from zero".

Examples:

Example 1. Calculate. $|12 − 4| \times 4 =$

Solution: First, solve $|12 − 4|$, $\rightarrow |12 − 4| = |8|$, the absolute value of 8 is 8, $|8| = 8$ Then: $8 \times 4 = 32$

Example 2. Solve. $\frac{|-16|}{4} \times |3 − 8| =$

Solution: First, find $|−16|$, $\rightarrow$ the absolute value of −16 is 16,
Then: $|−16| = 16$, $\frac{16}{4} \times |3 − 8| =$
Now, calculate $|3 − 8|$, $\rightarrow |3 − 8| = |−5|$, the absolute value of −5 is 5. $|−5| = 5$ then: $\frac{16}{4} \times 5 = 4 \times 5 = 20$

Example 3. Solve. $|9 − 3| \times \frac{|-3\times 8|}{6} =$

Solution: First, calculate $|9 − 3|$, $\rightarrow |9 − 3| = |6|$, the absolute value of 6 is 6, $|6| = 6$. Then: $6 \times \frac{|-3\times 8|}{6}$
Now calculate $|−3 \times 8|$, $\rightarrow |−3 \times 8| = |−24|$, the absolute value of −24 is 24, $|−24| = 24$ Then: $6 \times \frac{24}{6} = 6 \times 4 = 24$

CHAPTER 3: PRACTICES

✍ Find each sum or difference.

1) $13 - (-6) =$

2) $(-36) + 18 =$

3) $(-6) + (-22) =$

4) $54 + (-12) + 9 =$

5) $35 + (-24 + 4) =$

6) $(-17) + (-34 + 12) =$

7) $(-1) + (28 - 15) =$

8) $5 + (-9 + 12) =$

9) $(-10) + (-20) =$

10) $32 - (-5) =$

11) $(-9) + (24 - 3) =$

12) $8 - (-2 + 12) =$

13) $(-3) + (45 + 3) =$

14) $5 + (-30 + 6) =$

15) $(-6 + 1) + (-20) =$

16) $(-7) - (-20 + 2) =$

17) $(-6) - (2) =$

18) $(9 - 6) - (-3) =$

✍ Solve.

19) $4 \times (-8) =$

20) $(-27) \div (-9) =$

21) $(-2) \times (-9) \times 3 =$

22) $5 \times (-3) \times (-7) =$

23) $(-10 - 8) \div (-9) =$

24) $(-9 + 7) \times (-20) =$

25) $(-7) \times (-5) =$

26) $(-6) \times (-2 + 6) =$

27) $(-3) \times (-4) \times 3 =$

28) $(-8 - 2) \times (-1 + 4) =$

29) $(-9) \times (-20) =$

30) $6 \times (-2 + 9) =$

31) $(-5 - 6) \times (-2) =$

32) $(-4 - 2) \times (-3 - 7) =$

33) $(-9) \div (13 - 16) =$

34) $56 \times (-8) =$

35) $(-9 - 3) \div (-4) =$

36) $72 \div (-18 + 10) =$

✎ Evaluate each expression.

37) $2 + (6 \times 4) =$

38) $(7 \times 9) - 8 =$

39) $(-6) + (2 \times 9) =$

40) $(-2 - 4) + (3 \times 7) =$

41) $(28 \div 7) - (5 \times 3) =$

42) $(9 \times 3) + (6 \times 4) =$

43) $(36 \div 4) - (36 \div 6) =$

44) $(7 + 3) + (16 \div 2) =$

45) $(15 \times 3) - 16 =$

46) $8 - (7 \times 3) =$

47) $(9 + 15) \div (8 \div 4) =$

48) $2[(3 \times 6) + (16 \times 2)] =$

49) $(18 - 6) + (4 \times 2) =$

50) $2[(2 \times 3) - (8 \times 5)] =$

51) $(9 + 7) \div (16 \div 8) =$

52) $(3 + 9) \times (25 \times 2) =$

53) $3[(10 \times 9) \div (9 \times 5)] =$

54) $-6[(10 \times 9) \div (5 \times 6)] =$

✎ Find the answers.

55) $|-8| + |6 - 15| =$

56) $|-5 + 9| + |-3| =$

57) $|-6| + |2 - 10| =$

58) $|-8 + 3| - |4 - 8| =$

59) $|6 - 10| + |5 - 7| =$

60) $|-6| - |-9 - 19| + 5 =$

61) $|-9 + 2| - |3 - 5| + 6 =$

62) $4 + |3 - 7| + |2 - 6| =$

63) $\dfrac{|-6\ |}{8} \times \dfrac{|-4\ |}{6} =$

64) $\dfrac{|-3\ |}{6} \times \dfrac{|-56|}{8} =$

65) $\dfrac{|-72|}{9} \times \dfrac{|-45|}{5} =$

66) $|8 \times (-1)| \times \dfrac{|-24|}{3} =$

67) $|-2 \times 5| \times \dfrac{|-27|}{9} =$

68) $\dfrac{|-144|}{12} - |-6 \times 4| =$

69) $\dfrac{|-63|}{7} + |-9 \times 2| =$

70) $\dfrac{|-70|}{7} + |-8 \times 3| =$

71) $\dfrac{|-7 \times -3|}{7} \times \dfrac{|8 \times (-5)|}{8} =$

72) $\dfrac{|(-2) \times (-6)|}{4} \times \dfrac{|8 \times (-4)|}{2} =$

CHAPTER 3: ANSWERS

1) 19	25) 35	49) 20
2) −18	26) −24	50) −68
3) −28	27) 36	51) 8
4) 51	28) −30	52) 600
5) 15	29) 180	53) 6
6) −39	30) 42	54) −18
7) 12	31) 22	55) 17
8) 8	32) 80	56) 7
9) −30	33) 3	57) 14
10) 37	34) −448	58) 1
11) 12	35) 3	59) 6
12) −2	36) −9	60) −17
13) 45	37) 26	61) 11
14) −19	38) 55	62) 12
15) −25	39) 12	63) 64
16) 11	40) 15	64) 42
17) −8	41) −11	65) 72
18) 6	42) 51	66) 64
19) −32	43) 3	67) 30
20) 3	44) 18	68) −12
21) 54	45) 29	69) 27
22) 105	46) −13	70) 34
23) 2	47) 12	71) 15
24) 40	48) 100	72) 48

CHAPTER 4:

RATIOS AND PROPORTIONS

Math Topics that you'll learn in this chapter:

▶ Simplifying Ratios

▶ Proportional Ratios

▶ Similarity and Ratios

Simplifying Ratios

☑ Ratios are used to make comparisons between two numbers.

☑ Ratios can be written as a fraction, using the word "to", or with a colon. Example: $\frac{3}{4}$ or "3 to 4" or 3:4

☑ You can calculate equivalent ratios by multiplying or dividing both sides of the ratio by the same number.

Examples:

Example 1. Simplify. $9:3 =$

Solution: Both numbers 9 and 3 are divisible by 3 , $\Rightarrow 9 \div 3 = 3$, $3 \div 3 = 1$, Then: $9:3 = 3:1$

Example 2. Simplify. $\frac{24}{44} =$

Solution: Both numbers 24 and 44 are divisible by 4, $\Rightarrow$ $24 \div 4 = 6$, $44 \div 4 = 11$, Then: $\frac{24}{44} = \frac{6}{11}$

Example 3. There are 36 students in a class and 16 are girls. Write the ratio of girls to boys.

Solution: Subtract 16 from 36 to find the number of boys in the class. $36 - 16 = 20$. There are 20 boys in the class. So, the ratio of girls to boys is $16:20$. Now, simplify this ratio. Both 20 and 16 are divisible by 4. Then: $20 \div 4 = 5$, and $16 \div 4 = 4$. In the simplest form, this ratio is $4:5$

Example 4. A recipe calls for butter and sugar in the ratio $3:4$. If you're using 9 cups of butter, how many cups of sugar should you use?

Solution: Since you use 9 cups of butter, or 3 times as much, you need to multiply the amount of sugar by 3. Then: $4 \times 3 = 12$. So, you need to use 12 cups of sugar. You can solve this using equivalent fractions: $\frac{3}{4} = \frac{9}{12}$

PROPORTIONAL RATIOS

☑ Two ratios are proportional if they represent the same relationship.

☑ A proportion means that two ratios are equal. It can be written in two ways:
$$\frac{a}{b} = \frac{c}{d} \qquad a : b = c : d$$

☑ The proportion $\frac{a}{b} = \frac{c}{d}$ can be written as: $a \times d = c \times b$

Examples:

Example 1. Solve this proportion for x. $\quad \frac{3}{7} = \frac{12}{x}$

Solution: Use cross multiplication: $\frac{3}{7} = \frac{12}{x} \Rightarrow 3 \times x = 7 \times 12 \Rightarrow 3x = 84$

Divide both sides by 3 to find x: $\qquad x = \frac{84}{3} \Rightarrow x = 28$

Example 2. If a box contains red and blue balls in ratio of $3 : 7$ red to blue, how many red balls are there if 49 blue balls are in the box?

Solution: Write a proportion and solve. $\frac{3}{7} = \frac{x}{49}$

Use cross multiplication: $\quad 3 \times 49 = 7 \times x \Rightarrow 147 = 7x$

Divide to find x: $x = \frac{147}{7} \Rightarrow x = 21$. There are 21 red balls in the box.

Example 3. Solve this proportion for x. $\quad \frac{2}{9} = \frac{12}{x}$

Solution: Use cross multiplication: $\frac{2}{9} = \frac{12}{x} \Rightarrow 2 \times x = 9 \times 12 \Rightarrow 2x = 108$

Divide to find x: $x = \frac{108}{2} \Rightarrow x = 54$

Example 4. Solve this proportion for x. $\frac{6}{7} = \frac{18}{x}$

Solution: Use cross multiplication: $\frac{6}{7} = \frac{18}{x} \Rightarrow 6 \times x = 7 \times 18 \Rightarrow 6x = 126$

Divide to find x: $x = \frac{126}{6} \Rightarrow x = 21$

SIMILARITY AND RATIOS

☑ Two figures are similar if they have the same shape.

☑ Two or more figures are similar if the corresponding angles are equal, and the corresponding sides are in proportion.

Examples:

Example 1. The following triangles are similar. What is the value of the unknown side?

Solution: Find the corresponding sides and write a proportion.

$\frac{5}{10} = \frac{4}{x}$. Now, use the cross product to solve for x:

$\frac{5}{10} = \frac{4}{x} \rightarrow 5 \times x = 10 \times 4 \rightarrow 5x = 40$. Divide both sides by 5. Then: $5x = 40 \rightarrow \frac{5x}{5} = \frac{40}{5} \rightarrow x = 8$

The missing side is 8.

Example 2. Two rectangles are similar. The first is 6 feet wide and 20 feet long. The second is 15 feet wide. What is the length of the second rectangle?

Solution: Let's put x for the length of the second rectangle. Since two rectangles are similar, their corresponding sides are in proportion. Write a proportion and solve for the missing number.

$\frac{6}{15} = \frac{20}{x} \rightarrow 6x = 15 \times 20 \rightarrow 6x = 300 \rightarrow x = \frac{300}{6} = 50$

The length of the second rectangle is 50 feet.

CHAPTER 4: PRACTICES

✎ Reduce each ratio.

1) $3:21 = $ ___ : ___

2) $8:72 = $ ___ : ___

3) $21:49 = $ ___ : ___

4) $32:28 = $ ___ : ___

5) $35:45 = $ ___ : ___

6) $72:81 = $ ___ : ___

7) $36:54 = $ ___ : ___

8) $56:64 = $ ___ : ___

9) $12:36 = $ ___ : ___

10) $4:32 = $ ___ : ___

11) $16:48 = $ ___ : ___

12) $15:105 = $ ___ : ___

✎ Solve.

13) Bob has 18 red cards and 27 green cards. What is the ratio of Bob's red cards to his green cards? _____

14) In a party, 30 soft drinks are required for every 18 guests. If there are 240 guests, how many soft drinks are required? _____

15) Sara has 72 blue pens and 36 black pens. What is the ratio of Sara's black pens to her blue pens? _____

16) In Jack's class, 45 of the students are tall and 18 are short. In Michael's class 27 students are tall and 12 students are short. Which class has a higher ratio of tall to short students? _____

17) The price of 3 apples at the Quick Market is $1.44. The price of 5 of the same apples at Walmart is $2.45. Which place is the better buy? _____

18) The bakers at a Bakery can make 160 bagels in 4 hours. How many bagels can they bake in 14 hours? What is that rate per hour? _____

19) You can buy 5 cans of green beans at a supermarket for $3.40. How much does it cost to buy 35 cans of green beans? _____

✍ Solve each proportion.

20) $\frac{3}{4} = \frac{15}{x}$, $x =$

21) $\frac{9}{6} = \frac{x}{4}$, $x =$ _____

22) $\frac{3}{15} = \frac{2}{x}$, $x =$ _____

23) $\frac{5}{15} = \frac{3}{x}$, $x =$ _____

24) $\frac{24}{3} = \frac{x}{2}$, $x =$ _____

25) $\frac{8}{12} = \frac{10}{x}$, $x =$ _____

26) $\frac{3}{x} = \frac{2}{14}$, $x =$ _____

27) $\frac{10}{x} = \frac{3}{6}$, $x =$ _____

28) $\frac{15}{6} = \frac{x}{4}$, $x =$ _____

29) $\frac{x}{12} = \frac{5}{10}$, $x =$ _____

30) $\frac{18}{6} = \frac{3}{x}$, $x =$ _____

31) $\frac{3}{4} = \frac{24}{x}$, $x =$ _____

32) $\frac{8}{4} = \frac{x}{2}$, $x =$ _____

33) $\frac{12}{3} = \frac{x}{4}$, $x =$ _____

34) $\frac{24}{8} = \frac{x}{2}$, $x =$ _____

35) $\frac{5}{3} = \frac{x}{6}$, $x =$ _____

36) $\frac{10}{8} = \frac{x}{4}$, $x =$ _____

37) $\frac{x}{6} = \frac{6}{4}$, $x =$ _____

38) $\frac{x}{4} = \frac{7}{2}$, $x =$ _____

39) $\frac{9}{x} = \frac{3}{4}$, $x =$ _____

40) $\frac{10}{x} = \frac{1}{5}$, $x =$ _____

41) $\frac{9}{2} = \frac{x}{8}$, $x =$ _____

✍ Solve each problem.

42) Two rectangles are similar. The first is 6 *feet* wide and 24 *feet* long. The second is 10 *feet* wide. What is the length of the second rectangle?

_____

43) Two rectangles are similar. One is 4.8 *meters* by 6 *meters*. The longer side of the second rectangle is 27 *meters*. What is the other side of the second rectangle? _____

CHAPTER 4: ANSWERS

1) $1:7$

2) $1:9$

3) $3:7$

4) $8:7$

5) $7:9$

6) $8:9$

7) $2:3$

8) $7:8$

9) $1:3$

10) $1:8$

11) $1:3$

12) $1:7$

13) $2:3$

14) 144

15) $1:2$

16) $jack's\ class = \frac{45}{18} = \frac{5}{2}$

$Michael's\ class = \frac{27}{12} = \frac{9}{4}$

Jack's class has a higher ratio of tall to short student

17) Quick Market

18) 560

19) $\$23.80$

20) 20

21) 6

22) 10

23) 9

24) 16

25) 15

26) 21

27) 20

28) 10

29) 6

30) 1

31) 32

32) 16

33) 16

34) 6

35) 1

36) 5

37) 9

38) 14

39) 12

40) 50

41) 36

42) 40

43) $21.6\ meters$

CHAPTER 5:

PERCENTAGE

Math Topics that you'll learn in this chapter:

▶ Percentage Calculations

▶ Percent Problems

▶ Percent of Increase and Decrease

▶ Discount, Tax and Tip

▶ Simple Interest

PERCENT PROBLEMS

☑ Percent is a ratio of a number and 100. It always has the same denominator, 100. The percent symbol is "%".

☑ Percent means "per 100". So, 20% is 20/100.

☑ In each percent problem, we are looking for the base, or part or the percent.

☑ Use these equations to find each missing section in a percent problem:

❖ Base = Part ÷ Percent

❖ Part = Percent × Base

❖ Percent = Part ÷ Base

Examples:

Example 1. What is 25% of 60?

Solution: In this problem, we have percent (25%) and base (60) and we are looking for the "part". Use this formula: *part = percent × base*.
Then: $part = 25\% \times 60 = \frac{25}{100} \times 60 = 0.25 \times 60 = 15$. The answer:
25% of 60 is 15.

Example 2. 20 is what percent of 400?

Solution: In this problem, we are looking for the percent. Use this equation: *Percent = Part ÷ Base → Percent = 20 ÷ 400 = 0.05 = 5%*.
Then: 20 is 5 percent of 400.

PERCENT OF INCREASE AND DECREASE

☑ Percent of change (increase or decrease) is a mathematical concept that represents the degree of change over time.

☑ To find the percentage of increase or decrease:

    1. New Number – Original Number

    2. The result ÷ Original Number × 100

☑ Or use this formula: Percent of change $= \frac{new\ number - original\ number}{original\ number} \times 100$

☑ Note: If your answer is a negative number, then this is a percentage decrease. If it is positive, then this is a percentage increase.

Examples:

Example 1. The price of a shirt increases from \$20 to \$30. What is the percentage increase?

Solution: First, find the difference: $30 - 20 = 10$

Then: $10 \div 20 \times 100 = \frac{10}{20} \times 100 = 50$. The percentage increase is 50. It means that the price of the shirt increased by 50%.

Example 2. The price of a table increased from \$25 to \$40. What is the percent of increase?

Solution: Use percentage formula:

$Percent\ of\ change = \frac{new\ number - original\ number}{original\ number} \times 100 =$

$\frac{40-25}{25} \times 100 = \frac{15}{25} \times 100 = 0.6 \times 100 = 60$. The percentage increase is 60. It means that the price of the table increased by 60%.

DISCOUNT, TAX AND TIP

☑ To find the discount: Multiply the regular price by the rate of discount

☑ To find the selling price: Original price – discount

☑ To find tax: Multiply the tax rate to the taxable amount (income, property value, etc.)

☑ To find the tip, multiply the rate to the selling price.

Examples:

Example 1. With an 10% discount, Ella saved $45 on a dress. What was the original price of the dress?

Solution: let x be the original price of the dress. Then: $10\% \; of \; x = 45$. Write an equation and solve for x: $0.10 \times x = 45 \rightarrow x = \frac{45}{0.10} = 450$. The original price of the dress was $450.

Example 2. Sophia purchased a new computer for a price of $950 at the Apple Store. What is the total amount her credit card is charged if the sales tax is 7%?

Solution: The taxable amount is $950, and the tax rate is 7%. Then: $Tax = 0.07 \times 950 = 66.50$
$Final \; price = Selling \; price + Tax \rightarrow final \; price = \$950 + \$66.50 = \$1,016.50$

Example 3. Nicole and her friends went out to eat at a restaurant. If their bill was $80.00 and they gave their server a 15% tip, how much did they pay altogether?

Solution: First, find the tip. To find the tip, multiply the rate to the bill amount. $Tip = 80 \times 0.15 = 12$. The final price is: $80 + $12 = $92

SIMPLE INTEREST

☑ Simple Interest: The charge for borrowing money or the return for lending it.

☑ Simple interest is calculated on the initial amount (principal).

☑ To solve a simple interest problem, use this formula:

Interest = principal x rate x time $(I = p \times r \times t = prt)$

Examples:

Example 1. Find simple interest for $300 investment at 6% for 5 years.

Solution: Use Interest formula:
$I = prt$ ($P = \$300$, r $= 6\% = \frac{6}{100} = 0.06$ and $t = 5$)
Then: $I = 300 \times 0.06 \times 5 = \$90$

Example 2. Find simple interest for $1,600 at 5% for 2 years.

Solution: Use Interest formula:
$I = prt$ ($P = \$1,600$, r $= 5\% = \frac{5}{100} = 0.05$ and $t = 2$)
Then: $I = 1,600 \times 0.05 \times 2 = \$160$

Example 3. Andy received a student loan to pay for his educational expenses this year. What is the interest on the loan if he borrowed $6,500 at 8% for 6 years?

Solution: Use Interest formula:$I = prt$. $P = \$6,500$, r $= 8\% = 0.08$ and $t = 6$
Then: $I = 6,500 \times 0.08 \times 8 = \$3,120$

Example 4. Bob is starting his own small business. He borrowed $10,000 from the bank at a 6% rate for 6 months. Find the interest Bob will pay on this loan.

Solution: Use Interest formula:
$I = prt$. $P = \$10,000$, r $= 6\% = 0.06$ and $t = 0.5$ (6 months is half year).
Then: $I = 10,000 \times 0.06 \times 0.5 = \$300$

Chapter 5: Practices

✍ Solve each problem.

1) 10 is what percent of 80? ____%

2) 12 is what percent of 60? ____%

3) 20 is what percent of 80? ____%

4) 18 is what percent of 72? ____%

5) 16 is what percent of 50? ____%

6) 35 is what percent of 140? ____%

7) 12 is what percent of 240? ____%

8) 80 is what percent of 400? ____%

9) 60 is what percent of 300? ____%

10) 100 is what percent of 250? ____%

11) 25 is what percent of 400? ____%

12) 60 is what percent of 480? ____%

✍ Solve each problem.

13) Bob got a raise, and his hourly wage increased from $16 to $20. What is the percent increase? _____ %

14) The price of a pair of shoes increases from $30 to $36. What is the percent increase? ___ %

15) At a coffeeshop, the price of a cup of coffee increased from $1.30 to $1.56. What is the percent increase in the cost of the coffee? _____ %

16) A $40 shirt now selling for $28 is discounted by what percent? _____ %

17) Joe scored 20 out of 25 marks in Algebra, 30 out of 40 marks in science and 68 out of 80 marks in mathematics. In which subject his percentage of marks is best? _____

18) Emma purchased a computer for $408. The computer is regularly priced at $480. What was the percent discount Emma received on the computer? _____

19) A chemical solution contains 12% alcohol. If there is 42 ml of alcohol, what is the volume of the solution? _____

✍ Find the selling price of each item.

20) Original price of a computer: $700

Tax: 9%, Selling price: $_____

21) Original price of a laptop: $460

Tax: 20%, Selling price: $_____

22) Nicolas hired a moving company. The company charged $600 for its services, and Nicolas gives the movers a 12% tip. How much does Nicolas tip the movers? $_____

23) Mason has lunch at a restaurant and the cost of his meal is $50. Mason wants to leave a 25% tip. What is Mason's total bill, including tip? $_____

✍ Determine the simple interest for the following loans.

24) $480 at 6% for 5 $years.$ $___

25) $500 at 5% for 3 $years.$ $___

26) $360 at 3.5% for 2 $years.$ $___

27) $600 at 4% for 4 years. $___

✍ Solve.

28) A new car, valued at $25,000, depreciates at 7% per year. What is the value of the car one year after purchase? $_____

29) Sara puts $6,000 into an investment yielding 4% annual simple interest; she left the money in for five years. How much interest does Sara get at the end of those five years? $_____

CHAPTER 5: ANSWERS

1) 12.5%

2) 20%

3) 25%

4) 25%

5) 32%

6) 25%

7) 5%

8) 20%

9) 20%

10) 40%

11) 6.25%

12) 12.5%

13) 25%

14) 20%

15) 20%

16) 30%

17) Mathematics

18) 15%

19) 350

20) $763

21) $552

22) $72

23) $62.50

24) $144

25) $75

26) $25.20

27) $96

28) $23,250

29) $1200

CHAPTER 6:

EXPRESSIONS AND VARIABLES

Math Topics that you'll learn in this chapter:

▶ Simplifying Variable Expressions

▶ Simplifying Polynomial Expressions

▶ The Distributive Property

▶ Evaluating One Variable

▶ Evaluating Two Variables

SIMPLIFYING VARIABLE EXPRESSIONS

☑ In algebra, a variable is a letter used to stand for a number. The most common letters are $x, y, z, a, b, c, m, and\ n$.

☑ An algebraic expression is an expression that contains integers, variables, and math operations such as addition, subtraction, multiplication, division, etc.

☑ In an expression, we can combine "like" terms. (values with same variable and same power)

Examples:

Example 1. Simplify. $(2x + 3x + 4) =$

Solution: In this expression, there are three terms: $2x, 3x$, and 4. Two terms are "like terms": $2x$ and $3x$. Combine like terms. $2x + 3x = 5x$. Then: $(2x + 3x + 4) = 5x + 4$ (**remember you cannot combine variables and numbers.**)

Example 2. Simplify. $12 - 3x^2 + 5x + 4x^2 =$

Solution: Combine "like" terms: $-3x^2 + 4x^2 = x^2$.
Then: $12 - 3x^2 + 5x + 4x^2 = 12 + x^2 + 5x$. Write in standard form (biggest powers first): $12 + x^2 + 5x = x^2 + 5x + 12$

Example 3. Simplify. $(10x^2 + 2x^2 + 3x) =$

Solution: Combine like terms. Then: $(10x^2 + 2x^2 + 3x) = 12x^2 + 3x$

Example 4. Simplify. $15x - 3x^2 + 9x + 5x^2 =$

Solution: Combine "like" terms: $15x + 9x = 24x$, and $-3x^2 + 5x^2 = 2x^2$
Then: $15x - 3x^2 + 9x + 5x^2 = 24x + 2x^2$. Write in standard form (biggest powers first): $24x + 2x^2 = 2x^2 + 24x$

SIMPLIFYING POLYNOMIAL EXPRESSIONS

☑ In mathematics, a polynomial is an expression consisting of variables and coefficients that involves only the operations of addition, subtraction, multiplication, and non–negative integer exponents of variables. $P(x) = a_n x^n + a_{n-1} x^{n-1} + \ldots + a_2 x^2 + a_1 x + a_0$

☑ Polynomials must always be simplified as much as possible. It means you must add together any like terms. (values with same variable and same power)

Examples:

Example 1. Simplify this Polynomial Expressions. $x^2 - 5x^3 + 2x^4 - 4x^3$

Solution: Combine "like" terms: $-5x^3 - 4x^3 = -9x^3$
Then: $x^2 - 5x^3 + 2x^4 - 4x^3 = x^2 - 9x^3 + 2x^4$
Now, write the expression in standard form: $2x^4 - 9x^3 + x^2$

Example 2. Simplify this expression. $(2x^2 - x^3) - (x^3 - 4x^2) =$

Solution: First, use distributive property: →   multiply (−) into $(x^3 - 4x^2)$
$(2x^2 - x^3) - (x^3 - 4x^2) = 2x^2 - x^3 - x^3 + 4x^2$
Then combine "like" terms: $2x^2 - x^3 - x^3 + 4x^2 = 6x^2 - 2x^3$
And write in standard form: $6x^2 - 2x^3 = -2x^3 + 6x^2$

Example 3. Simplify. $4x^4 - 5x^3 + 15x^4 - 12x^3 =$

Solution: Combine "like" terms:
$-5x^3 - 12x^3 = -17x^3$ and $4x^4 + 15x^4 = 19x^4$
Then: $4x^4 - 5x^3 + 15x^4 - 12x^3 = 19x^4 - 17x^3$

The Distributive Property

☑ The distributive property (or the distributive property of multiplication over addition and subtraction) simplifies and solves expressions in the form of: $a(b + c)$ or $a(b - c)$

☑ The distributive property is multiplying a term outside the parentheses by the terms inside.

☑ Distributive Property rule: $a(b + c) = ab + ac$

Examples:

Example 1. Simply using the distributive property. $(-4)(x - 5)$

Solution: Use Distributive Property rule: $a(b + c) = ab + ac$
$(-4)(x - 5) = (-4 \times x) + (-4) \times (-5) = -4x + 20$

Example 2. Simply. $(3)(2x - 4)$

Solution: Use Distributive Property rule: $a(b + c) = ab + ac$
$(3)(2x - 4) = (3 \times 2x) + (3) \times (-4) = 6x - 12$

Example 3. Simply. $(-3)(3x - 5) + 4x$

Solution: First, simplify $(-3)(3x - 5)$ using the distributive property.
Then: $(-3)(3x - 5) = -9x + 15$
Now combine like terms: $(-3)(3x - 5) + 4x = -9x + 15 + 4x$
In this expression, $-9x$ and $4x$ are "like terms" and we can combine them.
$-9x + 4x = -5x$. Then: $-9x + 15 + 4x = -5x + 15$

EVALUATING ONE VARIABLE

☑ To evaluate one variable expression, find the variable and substitute a number for that variable.

☑ Perform the arithmetic operations.

Examples:

Example 1. Calculate this expression for x = 3. $15 - 3x$

Solution: First, substitute 3 for x
Then: $15 - 3x = 15 - 3(3)$
Now, use order of operation to find the answer: $15 - 3(3) = 15 - 9 = 6$

Example 2. Evaluate this expression for x = 1. $5x - 12$

Solution: First, substitute 1 for x,
Then: $5x - 12 = 5(1) - 12$
Now, use order of operation to find the answer: $5(1) - 12 = 5 - 12 = -7$

Example 3. Find the value of this expression when x = 5. $25 - 4x$

Solution: First, substitute 5 for x,
Then: $25 - 4x = 25 - 4(5) = 25 - 20 = 5$

Example 4. Solve this expression for $x = -2$. $12 + 3x$

Solution: Substitute -2 for x,
Then: $12 + 3x = 12 + 3(-2) = 12 - 6 = 6$

Evaluating Two Variables

☑ To evaluate an algebraic expression, substitute a number for each variable.

☑ Perform the arithmetic operations to find the value of the expression.

Examples:

Example 1. Calculate this expression for a = 3 and $b = -2$. $3a - 6b$

Solution: First, substitute 3 for a, and -2 for b ,

Then: $3a - 6b = 3(3) - 6(-2)$

Now, use order of operation to find the answer: $3(3) - 6(-2) = 9 + 12 = 21$

Example 2. Evaluate this expression for x = 3 and **y** = 1. $3x + 5y$

Solution: Substitute 3 for x, and 1 for y ,

Then: $3x + 5y = 3(3) + 5(1) = 9 + 5 = 14$

Example 3. Find the value of this expression $5(3a - 2b)$ when $a = 1$ and $b = 2$.

Solution: Substitute 1 for a, and 2 for b ,

Then: $5(3a - 2b) = 15a - 10b = 15(1) - 10(2) = 15 - 20 = -5$

Example 4. Solve this expression. $4x - 3y$, $x = 3$, $y = 5$

Solution: Substitute 3 for x, and 5 for y and simplify.

Then: $4x - 3y = 4(3) - 3(5) = 12 - 15 = -3$

CHAPTER 6: PRACTICES

✍ Simplify each expression.

1) $(9x - 5x - 7 + 4) =$

2) $(-12x - 7x + 6 - 3) =$

3) $(24x - 10x - 3) =$

4) $(-10x + 23x - 6) =$

5) $(32x + 8 - 20x - 4) =$

6) $3 + 6x^2 - 6 =$

7) $3x + 6x^4 - 6x =$

8) $-1 - 2x^2 - 8 =$

9) $67 + 6x - 1 - 9 =$

10) $3x^2 + 9x - 11x - 2 =$

11) $-3x^2 - 5x - 7x + 6 - 7 =$

12) $9x - 2x^2 + 8x =$

13) $12x^2 + 6x - 3x^2 + 12 =$

14) $10x^2 - 8x - 5x^2 + 4 =$

✍ Simplify each polynomial.

15) $8x^2 + 2x^3 - 4x^2 + 10x =$ --

16) $6x^4 + 3x^5 - 9x^4 + 7x^2 =$ --

17) $10x^3 + 12x - 3x^2 - 7x^3 =$ --

18) $(6x^3 - 2x^2) + (4x^2 - 14x) =$ --

19) $(13x^4 + 5x^3) + (2x^3 - 6x^4) =$ --

20) $(14x^5 - 9x^3) - (3x^3 + x^2) =$ --

21) $(10x^4 + 6x^3) - (x^3 - 65) =$ --

22) $(26x^4 + 5x^3) - (15x^3 - 3x^4) =$ --

23) $(10x^2 + 8x^3) + (25x^2 + 4x^3) =$ --

24) $(8x^4 - 3x^3) + (4x^3 - 7x^4) =$ --

✎ Use the distributive property to simply each expression.

25) $3(6 + 9x) =$ _____

26) $6(4 - 3x) =$ _____

27) $(-8)(3 - 4x) =$ _____

28) $(2 - 5x)(-6) =$ _____

29) $3(7 - 2x) =$

30) $(-x + 1)(-9) =$ _____

31) $(-3)(9x - 5) =$ _____

32) $(2x + 10)6 =$ _____

33) $(-1)(1 - 3x) =$ _____

34) $(6x - 1)(-9) =$ _____

✎ Evaluate each expression using the value given.

35) $x = -5,\quad 14 - x =$ ____

36) $x = -7,\quad x + 10 =$ ____

37) $x = 4,\quad 6x - 8 =$ ____

38) $x = 3,\quad 9 - 3x =$ ____

39) $x = -8,\quad 6x - 9 =$ ____

40) $x = 7,\quad 18 - 3x =$ ____

41) $x = -1,\quad 14x - 3 =$ ____

42) $x = 5,\quad 10 - x =$ ____

43) $x = 2,\quad 28 - 5x =$ ____

44) $x = -9,\quad 100 - 6x =$ ____

45) $x = 10,\quad 50 - 8x =$ ____

46) $x = 3,\quad 61x - 3 =$ ____

47) $x = 3,\quad 25x - 2 =$ ____

48) $x = -1,\quad 13 - 4x =$ ____

✎ Evaluate each expression using the values given.

49) $x = 2, y = -1,\quad 3x - 6y =$ _____

50) $a = 3, b = 6,\quad 7a + 2b =$ _____

51) $x = 3, y = 2,\quad 5x - 23y + 9 =$ _____

52) $a = 7, b = 5,\quad -9a + 3b + 8 =$ _____

53) $x = 3, y = 6,\quad 3x + 15 + 6y =$ _____

CHAPTER 6: ANSWERS

1) $4x - 3$

2) $-19x + 3$

3) $14x - 3$

4) $13x - 6$

5) $12x - 4$

6) $6x^2 - 3$

7) $6x^4 - 3x$

8) $-2x^2 - 9$

9) $6x + 57$

10) $10x^2 - 2x - 2$

11) $-3x^2 - 12x - 1$

12) $-2x^2 + 17x$

13) $9x^2 + 6x + 12$

14) $5x^2 - 8x + 4$

15) $2x^3 + 4x^2 + 10x$

16) $3x^5 - 3x^4 + 7x$

17) $3x^3 - 3x^2 + 12x$

18) $6x^3 + 2x^2 - 14x$

19) $7x^4 + 7x^3$

20) $14x^5 - 12x^3 - x^2$

21) $10x^4 + 5x^3 + 65$

22) $29x^4 - 10x^3$

23) $12x^3 + 35x^2$

24) $x^4 + x^3$

25) $27x + 18$

26) $-18x + 24$

27) $32x - 24$

28) $30x - 12$

29) $-6x + 21$

30) $9x - 9$

31) $-27x + 15$

32) $12x + 60$

33) $3x - 1$

34) $-54x + 9$

35) 19

36) 3

37) 16

38) 0

39) -57

40) -3

41) -17

42) 5

43) 18

44) 154

45) -30

46) 180

47) 73

48) 17

49) 12

50) 33

51) -22

52) -40

53) 60

CHAPTER 7:

EQUATIONS AND INEQUALITIES

Math Topics that you'll learn in this chapter:

▶ One-Step Equations

▶ Multi-Step Equations

▶ System of Equations

▶ Graphing Single–Variable Inequalities

▶ One-Step Inequalities

▶ Multi-Step Inequalities

ONE–STEP EQUATIONS

☑ The values of two expressions on both sides of an equation are equal. Example: $ax = b$. In this equation, ax is equal to b.

☑ Solving an equation means finding the value of the variable.

☑ You only need to perform one Math operation to solve the one-step equations.

☑ To solve a one-step equation, find the inverse (opposite) operation is being performed.

☑ The inverse operations are:

❖ Addition and subtraction

❖ Multiplication and division

Examples:

Example 1. Solve this equation for x. $3x = 18, x = ?$

Solution: Here, the operation is multiplication (variable x is multiplied by 3) and its inverse operation is division. To solve this equation, divide both sides of equation by 3: $3x = 18 \rightarrow \frac{3x}{3} = \frac{18}{3} \rightarrow x = 6$

Example 2. Solve this equation. $x + 15 = 0$, $x = ?$

Solution: In this equation 15 is added to the variable x. The inverse operation of addition is subtraction. To solve this equation, subtract 15 from both sides of the equation: $x + 15 - 15 = 0 - 15$. Then: $\rightarrow x = -15$

Example 3. Solve this equation for x. $x + 23 = 0$

Solution: Here, the operation is subtraction and its inverse operation is addition. To solve this equation, add 23 to both sides of the equation: $x + 23 - 23 = 0 - 23 \rightarrow x = -23$

MULTI–STEP EQUATIONS

☑ To solve a multi-step equation, combine "like" terms on one side.

☑ Bring variables to one side by adding or subtracting.

☑ Simplify using the inverse of addition or subtraction.

☑ Simplify further by using the inverse of multiplication or division.

☑ Check your solution by plugging the value of the variable into the original equation.

Examples:

Example 1. Solve this equation for x. $3x + 6 = 16 - 2x$

Solution: First, bring variables to one side by adding $2x$ to both sides. Then: $3x + 6 = 16 - 2x \rightarrow 3x + 6 + 2x = 16 - 2x + 2x$.

Simplify: $5x + 6 = 16$ Now, subtract 6 from both sides of the equation:

$5x + 6 - 6 = 16 - 6 \rightarrow 5x = 10 \rightarrow$ Divide both sides by 5:

$5x = 10 \rightarrow \dfrac{5x}{5} = \dfrac{10}{5} \rightarrow x = 2$

Let's check this solution by substituting the value of 2 for x in the original equation:

$x = 2 \rightarrow 3x + 6 = 16 - 2x \rightarrow 3(2) + 6 = 16 - 2(2) \rightarrow 6 + 6 = 16 - 4 \rightarrow 12 = 12$

The answer $x = 2$ is correct.

Example 2. Solve this equation for x. $-4x + 4 = 16$

Solution: Subtract 4 from both sides of the equation.

$-4x + 4 - 4 = 16 - 4 \rightarrow -4x = 12$

Divide both sides by -4, then: $-4x = 12 \rightarrow \dfrac{-4x}{-4} = \dfrac{12}{-4} \rightarrow x = -3$

Now, check the solution:

$x = -3 \rightarrow -4x + 4 = 16 \rightarrow -4(-3) + 4 = 16 \rightarrow 16 = 16$

The answer $x = -2$ is correct.

SYSTEM OF EQUATIONS

☑ A system of equations contains two equations and two variables. For example, consider the system of equations: $x - y = 1, x + y = 5$

☑ The easiest way to solve a system of equations is using the elimination method. The elimination method uses the addition property of equality. You can add the same value to each side of an equation.

☑ For the first equation above, you can add $x + y$ to the left side and 5 to the right side of the first equation: $x - y + (x + y) = 1 + 5$. Now, if you simplify, you get: $x - y + (x + y) = 1 + 5 \rightarrow 2x = 6 \rightarrow x = 3$. Now, substitute 3 for the x in the first equation: $3 - y = 1$. By solving this equation, $y = 2$

Example:

What is the value of x + y in this system of equations?

$$\begin{cases} x + 2y = 6 \\ 2x - y = -8 \end{cases}$$

Solution: Solving a System of Equations by Elimination:
Multiply the first equation by (-2), then add it to the second equation.

$$\begin{matrix} -2(x + 2y = 6) \\ \underline{2x - y = -8} \end{matrix} \Rightarrow \begin{matrix} -2x - 4y = -12 \\ 2x - y = -8 \end{matrix} \Rightarrow -5y = -20 \Rightarrow y = 4$$

Plug in the value of y into one of the equations and solve for x.

$x + 2(4) = 6 \Rightarrow x + 8 = 6 \Rightarrow x = 6 - 8 \Rightarrow x = -2$

Thus, $x + y = -2 + 4 = 2$

GRAPHING SINGLE–VARIABLE INEQUALITIES

☑ An inequality compares two expressions using an inequality sign.

☑ Inequality signs are: "less than" <, "greater than" >, "less than or equal to" ≤, and "greater than or equal to" ≥.

☑ To graph a single–variable inequality, find the value of the inequality on the number line.

☑ For less than (<) or greater than (>) draw an open circle on the value of the variable. If there is an equal sign too, then use a filled circle.

☑ Draw an arrow to the right for greater or to the left for less than.

Examples:

Example 1. Draw a graph for this inequality. $x > 3$

Solution: Since the variable is greater than 3, then we need to find 3 in the number line and draw an open circle on it. Then, draw an arrow to the right.

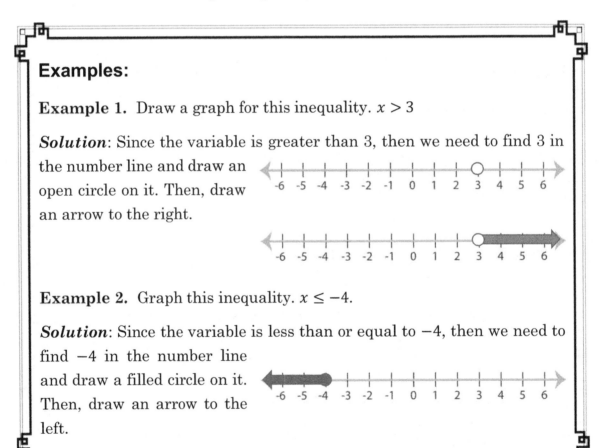

Example 2. Graph this inequality. $x \leq -4$.

Solution: Since the variable is less than or equal to −4, then we need to find −4 in the number line and draw a filled circle on it. Then, draw an arrow to the left.

ONE–STEP INEQUALITIES

☑ An inequality compares two expressions using an inequality sign.

☑ Inequality signs are: "less than" <, "greater than" >, "less than or equal to" ≤, and "greater than or equal to" ≥.

☑ You only need to perform one Math operation to solve the one-step inequalities.

☑ To solve one-step inequalities, find the inverse (opposite) operation is being performed.

☑ For dividing or multiplying both sides by negative numbers, flip the direction of the inequality sign.

Examples:

Example 1. Solve this inequality for x. $x + 3 \geq 4$

Solution: The inverse (opposite) operation of addition is subtraction. In this inequality, 3 is added to x. To isolate x we need to subtract 3 from both sides of the inequality.
Then: $x + 3 \geq 4 \rightarrow x + 3 - 3 \geq 4 - 3 \rightarrow x \geq 1$. The solution is: $x \geq 1$

Example 2. Solve the inequality. $x - 5 > -4$.

Solution: 5 is subtracted from x. Add 5 to both sides.
$x - 5 > -4 \rightarrow x - 5 + 5 > -4 + 5 \rightarrow x > 1$

Example 3. Solve. $2x \leq -4$.

Solution: 2 is multiplied to x. Divide both sides by 2.
Then: $2x \leq -4 \rightarrow \frac{2x}{2} \leq \frac{-4}{2} \rightarrow x \leq -2$

Example 4. Solve. $-6x \leq 12$.

Solution: -6 is multiplied to x. Divide both sides by -6. Remember when dividing or multiplying both sides of an inequality by negative numbers, flip the direction of the inequality sign.
Then: $-6x \leq 12 \rightarrow \frac{-6x}{-6} \geq \frac{12}{-6} \rightarrow x \geq -2$

MULTI–STEP INEQUALITIES

☑ To solve a multi-step inequality, combine "like" terms on one side.

☑ Bring variables to one side by adding or subtracting.

☑ Isolate the variable.

☑ Simplify using the inverse of addition or subtraction.

☑ Simplify further by using the inverse of multiplication or division.

☑ For dividing or multiplying both sides by negative numbers, flip the direction of the inequality sign.

Examples:

Example 1. Solve this inequality. $2x - 3 \leq 5$

Solution: In this inequality, 3 is subtracted from $2x$. The inverse of subtraction is addition. Add 3 to both sides of the inequality:
$2x - 3 + 3 \leq 5 + 3 \rightarrow 2x \leq 8$
Now, divide both sides by 2. Then: $2x \leq 8 \rightarrow \frac{2x}{2} \leq \frac{8}{2} \rightarrow x \leq 4$
The solution of this inequality is $x \leq 4$.

Example 2. Solve this inequality. $3x + 9 < 12$

Solution: First, subtract 9 from both sides: $3x + 9 - 9 < 12 - 9$
Then simplify: $3x + 9 - 9 < 12 - 9 \rightarrow 3x < 3$
Now divide both sides by 3: $\frac{3x}{3} < \frac{3}{3} \rightarrow x < 1$

Example 3. Solve this inequality. $-2x + 4 \geq 6$

Solution: First, subtract 4 from both sides:
$-2x + 4 - 4 \geq 6 - 4 \rightarrow -2x \geq 2$
Divide both sides by -2. Remember that you need to flip the direction of inequality sign. $-2x \geq 2 \rightarrow \frac{-2x}{-2} \leq \frac{2}{-2} \rightarrow x \leq -1$

CHAPTER 7: PRACTICES

✍ Solve each equation. (One–Step Equations)

1) $x + 8 = 4, x =$ _____

2) $3 = 12 - x, x =$ _____

3) $-4 = 9 + x, x =$ _____

4) $x - 6 = -9, x =$ _____

5) $18 = x + 8, x =$ _____

6) $15 - x = -4, x =$ _____

7) $25 - x = 8, x =$ _____

8) $6 + x = 27, x =$ _____

9) $10 - x = -8, x =$ _____

10) $36 - x = -5, x =$ _____

✍ Solve each equation. (Multi–Step Equations)

11) $6(x + 8) = 24, \ x =$ ____

12) $-9(9 - x) = 18, x =$ ____

13) $7 = -7(x + 3), x =$ ____

14) $-16 = 2(10 - 6x), x =$ ____

15) $6(x + 1) = -24, x =$ ____

16) $-3(7 + 9x) = 33, x =$ ____

17) $-7(5 - x) = 14, x =$ ____

18) $-1(3 - x) = 10, x =$ ____

✍ Solve each system of equations.

19) $\begin{cases} -2x + 2y = -4 \\ 4x - 9y = 28 \end{cases} \quad \begin{matrix} x = \\ y = \end{matrix}$

20) $\begin{cases} x + 8y = -5 \\ 2x + 6y = 0 \end{cases} \quad \begin{matrix} x = \\ y = \end{matrix}$

21) $\begin{cases} 4x - 3y = -2 \\ x - y = 3 \end{cases} \quad \begin{matrix} x = \\ y = \end{matrix}$

22) $\begin{cases} 2x + 9y = 17 \\ -3x + 8y = 39 \end{cases} \quad \begin{matrix} x = \\ y = \end{matrix}$

✍ Draw a graph for each inequality.

23) $x \leq -3$

24) $x > -5$

✍ Solve each inequality and graph it.

25) $x - 2 \geq -2$

26) $2x - 3 < 9$

✍ Solve each inequality.

27) $4x + 12 > -8$

28) $3x + 14 > 5$

29) $-16 + 3x \leq 20$

30) $-18 + 6x \leq -24$

31) $8 + 2x \leq 16$

32) $5(x + 2) \geq 6$

33) $2(3 + x) \geq 10$

34) $6x - 10 < 14$

35) $12x + 8 < 32$

36) $8(4 + x) \geq 16$

37) $2(x - 5) \geq 18$

38) $x + 10 < 3$

39) $2(x - 4) \geq 20$

40) $-8 + 9x > 28$

41) $-4 + 8x > 60$

42) $-2 + 7x > 40$

CHAPTER 7: ANSWERS

1) -4

2) 9

3) -13

4) -3

5) 10

6) 19

7) 17

8) 21

9) 18

10) 41

11) -4

12) 11

13) -4

14) 3

15) -5

16) -2

17) 7

18) 13

19) $x = -2, y = -4$

20) $x = 3, y = -1$

21) $x = -11, y = -14$

22) $x = -5, y = 3$

23) $x \le -3$

24) $x > -5$

25) $x \ge 0$

26) $x < 6$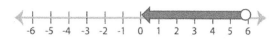

27) $x > -5$

28) $x > -3$

29) $x \le 12$

30) $x \le -1$

31) $x \le 4$

32) $x \ge -\frac{4}{5}$

33) $x \ge 2$

34) $x < 4$

35) $x < 2$

36) $x \ge -2$

37) $x \ge 14$

38) $x < -7$

39) $x \ge 14$

40) $x > 4$

41) $x > 8$

42) $x > 6$

CHAPTER 8:

LINES AND SLOPE

Math Topics that you'll learn in this chapter:

▶ Finding Slope

▶ Graphing Lines Using Slope–Intercept Form

▶ Writing Linear Equations

▶ Graphing Linear Inequalities

▶ Finding Midpoint

▶ Finding Distance of Two Points

FINDING SLOPE

✅ The slope of a line represents the direction of a line on the coordinate plane.

✅ A coordinate plane contains two perpendicular number lines. The horizontal line is x and the vertical line is y. The point at which the two axes intersect is called the origin. An ordered pair (x, y) shows the location of a point.

✅ A line on a coordinate plane can be drawn by connecting two points.

✅ To find the slope of a line, we need the equation of the line or two points on the line.

✅ The slope of a line with two points A (x_1, y_1) and B (x_2, y_2) can be found by using this formula: $\frac{y_2 - y_1}{x_2 - x_1} = \frac{rise}{run}$

✅ The equation of a line is typically written as $y = mx + b$ where m is the slope and b is the y-intercept.

Examples:

Example 1. Find the slope of the line through these two points:

A$(2, -7)$ and B$(4, 3)$.

Solution: Slope $= \frac{y_2 - y_1}{x_2 - x_1}$. Let (x_1, y_1) be A$(2, -7)$ and (x_2, y_2) be B$(4, 3)$.

(Remember, you can choose any point for (x_1, y_1) and (x_2, y_2)).

Then: slope $= \frac{y_2 - y_1}{x_2 - x_1} = \frac{3 - 7}{4 - 2} = \frac{10}{2} = 5$

The slope of the line through these two points is 5.

Example 2. Find the slope of the line with equation $y = 3x + 6$

Solution: when the equation of a line is written in the form of $y = mx + b$, the slope is m. In this line: $y = 3x + 6$, the slope is 3.

GRAPHING LINES USING SLOPE–INTERCEPT FORM

☑ Slope–intercept form of a line: given the slope **m** and the **y**-intercept (the intersection of the line and y-axis) **b**, then the equation of the line is:

$$y = mx + b$$

☑ To draw the graph of a linear equation in a slope-intercept form on the xy coordinate plane, find two points on the line by plugging two values for x and calculating the values of y.

☑ You can also use the slope (m) and one point to graph the line.

Example:

Example 1. Sketch the graph of y = 2x − 4.

Solution: To graph this line, we need to find two points. When x is zero the value of y is −4. And when x is 2 the value of y is 0.

$$x = 0 \rightarrow y = 2(0) - 4 = -4,$$
$$y = 0 \rightarrow 0 = 2x - 4 \rightarrow x = 2$$

Now, we have two points: $(0, -4)$ and $(2, 0)$.

Find the points on the coordinate plane and graph the line. Remember that the slope of the line is 2.

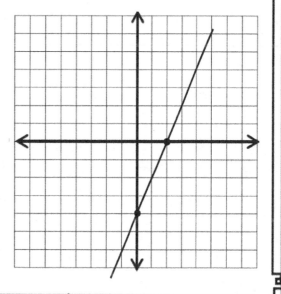

WRITING LINEAR EQUATIONS

☑ The equation of a line in slope-intercept form: $y = mx + b$

☑ To write the equation of a line, first identify the slope.

☑ Find the y-intercept. This can be done by substituting the slope and the coordinates of a point (x, y) on the line.

Examples:

Example 1. What is the equation of the line that passes through $(2, -4)$ and has a slope of 8?

Solution: The general slope-intercept form of the equation of a line is $y = mx + b$, where m is the slope and b is the y-intercept.

By substitution of the given point and given slope:

$y = mx + b \rightarrow -4 = (2)(8) + b$. So, $b = -4 - 16 = -20$, and the required equation is $y = 8x - 20$

Example 2. Write the equation of the line through two points $A(2, 1)$ and $B(-2, 5)$.

Solution: First, find the slope: $Slop = \frac{y_2 - y_1}{x_2 - x_1} = \frac{5 - 1}{-2 - 2} = \frac{4}{-4} = -1 \rightarrow m = -1$

To find the value of b, use either points and plug in the values of x and y in the equation. The answer will be the same: $y = -x + b$. Let's check both points. Then: $(2, 1) \rightarrow y = mx + b \rightarrow 1 = -1(2) + b \rightarrow b = 3$

$(-2, 5) \rightarrow y = mx + b \rightarrow 5 = -1(-2) + b \rightarrow b = 3$.

The y-intercept of the line is 3. The equation of the line is: $y = -x + 3$

Example 3. What is the equation of the line that passes through $(2, -1)$ and has a slope of 5?

Solution: The general slope-intercept form of the equation of a line is $y = mx + b$, where m is the slope and b is the y-intercept. By substitution of the given point and given slope:$y = mx + b \rightarrow -1 = (5)(2) + b$

So, $b = -1 - 10 = -11$, and the equation of the line is: $y = 5x - 11$.

FINDING MIDPOINT

☑ The middle of a line segment is its midpoint.

☑ The Midpoint of two endpoints A (x_1, y_1) and B (x_2, y_2) can be found using this formula: M $(\frac{x_1+x_2}{2}, \frac{y_1+y_2}{2})$

Examples:

Example 1. Find the midpoint of the line segment with the given endpoints. $(1, -3), (3, 7)$

Solution: Midpoint = $\left(\frac{x_1+x_2}{2}, \frac{y_1+y_2}{2}\right) \to (x_1, y_1) = (1, -3)$ and $(x_2, y_2) = (3, 7)$
Midpoint = $\left(\frac{1+3}{2}, \frac{-3+7}{2}\right) \to \left(\frac{4}{2}, \frac{4}{2}\right) \to M(2, 2)$

Example 2. Find the midpoint of the line segment with the given endpoints. $(-4, 5), (8, -7)$

Solution: Midpoint = $\left(\frac{x_1+x_2}{2}, \frac{y_1+y_2}{2}\right) \to (x_1, y_1) = (-4, 5)$ and $(x_2, y_2) = (8, -7)$
Midpoint = $\left(\frac{-4+8}{2}, \frac{5-7}{2}\right) \to \left(\frac{4}{2}, \frac{-2}{2}\right) \to M(2, -1)$

Example 3. Find the midpoint of the line segment with the given endpoints. $(5, -2), (1, 10)$

Solution: Midpoint = $\left(\frac{x_1+x_2}{2}, \frac{y_1+y_2}{2}\right) \to (x_1, y_1) = (5, -2)$ and $(x_2, y_2) = (1, 10)$
Midpoint = $\left(\frac{5+1}{2}, \frac{-2+10}{2}\right) \to \left(\frac{6}{2}, \frac{8}{2}\right) \to M(3, 4)$

Example 4. Find the midpoint of the line segment with the given endpoints. $(2, 3), (12, -9)$

Solution: Midpoint = $\left(\frac{x_1+x_2}{2}, \frac{y_1+y_2}{2}\right) \to (x_1, y_1) = (2, 3)$ and $(x_2, y_2) = (12, -3)$
Midpoint = $\left(\frac{2+12}{2}, \frac{3-9}{2}\right) \to \left(\frac{14}{2}, \frac{-6}{2}\right) \to M(7, -3)$

FINDING DISTANCE OF TWO POINTS

 Use the following formula to find the distance of two points with the coordinates A (x_1, y_1) and B (x_2, y_2):

$$d = \sqrt{(x_2 - x_1)^2 + (y_2 - y_1)^2}$$

Examples:

Example 1. Find the distance between $(4, 6)$ and $(1, 2)$.

Solution: Use distance of two points formula: $d = \sqrt{(x_2 - x_1)^2 + (y_2 - y_1)^2}$ $(x_1, y_1) = (4, 6)$ and $(x_2, y_2) = (1, 2)$. Then: $d = \sqrt{(x_2 - x_1)^2 + (y_2 - y_1)^2} \rightarrow$ $d = \sqrt{(1 - (4))^2 + (2 - 6)^2} = \sqrt{(-3)^2 + (-4)^2} = \sqrt{9 + 16} = \sqrt{25} = 5 \rightarrow d = 5$

Example 2. Find the distance of two points $(-6, -10)$ and $(-2, -10)$.

Solution: Use distance of two points formula: $d = \sqrt{(x_2 - x_1)^2 + (y_2 - y_1)^2}$ $(x_1, y_1) = (-6, -10)$, and $(x_2, y_2) = (-2, -10)$

Then: $d = \sqrt{(x_2 - x_1)^2 + (y_2 - y_1)^2} \rightarrow d = \sqrt{(-2 - (-6))^2 + (-10 - (-10))^2} =$ $\sqrt{(4)^2 + (0)^2} = \sqrt{16 + 0} = \sqrt{16} = 4$. Then: $d = 4$

Example 3. Find the distance between $(-6, 5)$ and $(-2, 2)$.

Solution: Use distance of two points formula: $d = \sqrt{(x_2 - x_1)^2 + (y_2 - y_1)^2}$ $(x_1, y_1) = (-6, 5)$ and $(x_2, y_2) = (-2, 2)$. Then: $d = \sqrt{(x_2 - x_1)^2 + (y_2 - y_1)^2}$ $d = \sqrt{(-2 - (-6))^2 + (2 - 5)^2} = \sqrt{(4)^2 + (-3)^2} = \sqrt{16 + 9} = \sqrt{25} = 5$

CHAPTER 8: PRACTICES

✍ Find the slope of each line.

1) $y = x - 3$

2) $y = -6x + 4$

3) $y = 3x - 9$

4) Line through $(-1, 3)$ and $(5, 0)$

5) Line through $(4, 0)$ and $(-2, 6)$

6) Line through $(-3, -6)$ and $(0, 3)$

✍ Sketch the graph of each line. (Using Slope–Intercept Form)

7) $y = x + 4$

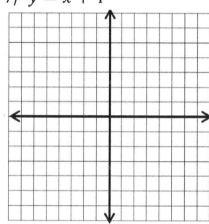

8) $y = 2x - 5$

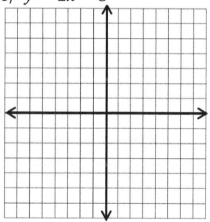

✍ Solve.

9) What is the equation of a line with slope 3 and intercept 18? _____

10) What is the equation of a line with slope 2 and passes through point $(2, 6)$?

_____

11) What is the equation of a line with slope -4 and passes through point $(-4, 8)$?

_____

12) The slope of a line is -2 and it passes through point $(-4, 3)$. What is the equation of the line? _____

13) The slope of a line is 5 and it passes through point $(-6, 3)$. What is the equation of the line? _____

✎ **Sketch the graph of each linear inequality.**

14) $y > 2x - 2$

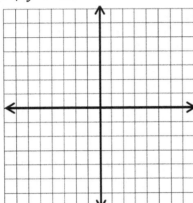

15) $y < -x + 3$

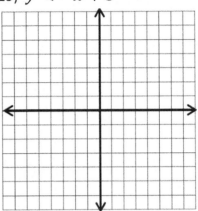

✎ **Find the midpoint of the line segment with the given endpoints.**

16) $(3, 6), (-1, 8)$

17) $(-2, 4), (8, 4)$

18) $(8, -3), (-2, 1)$

19) $(15, -9), (-3, 1)$

20) $(-10, 4), (6, 8)$

21) $(6, 12), (2, -4)$

22) $(4, 8), (-2, 0)$

23) $(0, 8), (-6, 6)$

✎ **Find the distance between each pair of points.**

24) $(-1, 6), (-5, 3)$

25) $(2, -2), (7, 10)$

26) $(-1, -4), (5, 4)$

27) $(6, -1), (-6, 8)$

28) $(2, -5), (-6, 10)$

29) $(0, 6), (4, 6)$

30) $(6, 3), (9, -1)$

31) $(0, -2), (10, 22)$

32) $(5, -6), (-11, 24)$

33) $(6, -10), (-6, 6)$

CHAPTER 8: ANSWERS

1) 1

2) -6

3) 3

4) $-\dfrac{1}{2}$

5) -1

6) 3

7) $y = x + 2$

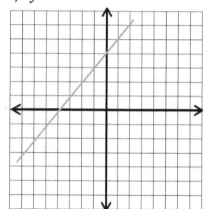

8) $y = 2x - 3$

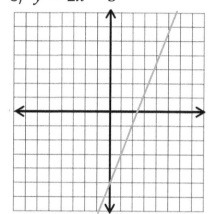

9) $y = 3x + 18$

10) $y = 2x + 2$

11) $y = -4x - 8$

12) $y = -2x - 5$

13) $y = 5x + 33$

14) $y > 2x - 2$

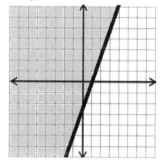

15) $y < -x + 3$

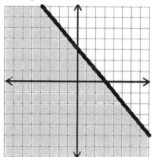

16) $(1, 7)$

17) $(3, 4)$

18) $(3, -1)$

19) $(6, -4)$

20) $(-2, 6)$

21) $(4, 4)$

22) $(1, 4)$

23) $(-3, 7)$

24) 5

25) 13

26) 10

27) 15

28) 17

29) 4

30) 5

31) 26

32) 34

33) 20

CHAPTER 9:

EXPONENTS AND VARIABLES

Math Topics that you'll learn in this chapter:

▶ Multiplication Property of Exponents

▶ Division Property of Exponents

▶ Powers of Products and Quotients

▶ Zero and Negative Exponents

▶ Negative Exponents and Negative Bases

▶ Scientific Notation

▶ Radicals

MULTIPLICATION PROPERTY OF EXPONENTS

☑ Exponents are shorthand for repeated multiplication of the same number by itself. For example, instead of 2×2, we can write 2^2. For $3 \times 3 \times 3 \times 3$, we can write 3^4

☑ In algebra, a variable is a letter used to stand for a number. The most common letters are: $x, y, z, a, b, c, m,$ and n.

☑ Exponent's rules: $x^a \times x^b = x^{a+b}$ $\frac{x^a}{x^b} = x^{a-b}$

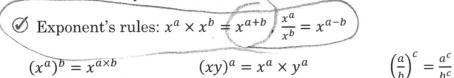

$$(x^a)^b = x^{a \times b} \qquad\qquad (xy)^a = x^a \times y^a \qquad\qquad \left(\frac{a}{b}\right)^c = \frac{a^c}{b^c}$$

Examples:

Example 1. Multiply. $4x^3 \times 2x^2$

Solution: Use Exponent's rules: $x^a \times x^b = x^{a+b} \rightarrow x^3 \times x^2 = x^{3+2} = x^5$
Then: $4x^3 \times 2x^2 = 8x^5$

Example 2. Simplify. $\left(x^3 y^5\right)^2$

Solution: Use Exponent's rules: $(x^a)^b = x^{a \times b}$.
Then: $\left(x^3 y^5\right)^2 = x^{3 \times 2} y^{5 \times 2} = x^6 y^{10}$

Example 3. Multiply. $-2x^5 \times 7x^3$

Solution: Use Exponent's rules: $x^a \times x^b = x^{a+b} \rightarrow x^5 \times x^3 = x^{5+3} = x^8$
Then: $-2x^5 \times 7x^3 = -14x^8$

Example 4. Simplify. $(x^2 y^4)^3$

Solution: Use Exponent's rules: $(x^a)^b = x^{a \times b}$.
Then: $(x^2 y^4)^3 = x^{2 \times 3} y^{4 \times 3} = x^6 y^{12}$

DIVISION PROPERTY OF EXPONENTS

☑ Exponents are shorthand for repeated multiplication of the same number by itself. For example, instead of 3×3, we can write 3^2. For $2 \times 2 \times 2$, we can write 2^3

☑ For division of exponents use following formulas:

$$\frac{x^a}{x^b} = x^{a-b} , x \neq 0, \quad \frac{x^a}{x^b} = \frac{1}{x^{b-a}} , x \neq 0, \qquad \frac{1}{x^b} = x^{-b}$$

Examples:

Example 1. Simplify. $\frac{12x^2y}{4xy^3} =$

Solution: First, cancel the common factor: $4 \to \frac{12x^2y}{4xy^3} = \frac{3x^2y}{xy^3}$

Use Exponent's rules: $\frac{x^a}{x^b} = x^{a-b} \to \frac{x^2}{x} = x^{2-1} = x$ and $\frac{y}{y^3} = \frac{1}{y^{3-1}} = \frac{1}{y^2}$

Then: $\frac{12x^2y}{4xy^3} = \frac{3x}{y^2}$

Example 2. Simplify. $\frac{18x^6}{2x^3} =$

Solution: Use Exponent's rules: $\frac{x^a}{x^b} = x^{b-a} \to \frac{x^6}{x^3} = x^{6-3} = x^3$

Then: $\frac{18x^6}{2x^3} = 9x^3$

Example 3. Simplify. $\frac{8x^3y}{40x^2y^3} =$ $\frac{2x}{10y^2}$ $\frac{x}{5y^2}$

Solution: First, cancel the common factor: $8 \to \frac{8x^3y}{40x^2y^3} = \frac{x^3y}{5x^2y^3}$

Use Exponent's rules: $\frac{x^a}{x^b} = x^{a-b} \to \frac{x^3}{x^2} = x^{3-2} = x$

Then: $\frac{8x^3y}{40x^2y^3} = \frac{xy}{5y^3} \to$ now cancel the common factor: $y \to \frac{xy}{5y^3} = \frac{x}{5y^2}$

POWERS OF PRODUCTS AND QUOTIENTS

☑ Exponents are shorthand for repeated multiplication of the same number by itself. For example, instead of $2 \times 2 \times 2$, we can write 2^3. For $3 \times 3 \times 3 \times 3$, we can write 3^4

☑ For any nonzero numbers a and b and any integer x, $(ab)^x = a^x \times b^x$ and $\left(\dfrac{a}{b}\right)^c = \dfrac{a^c}{b^c}$

Examples:

Example 1. Simplify. $(6x^2y^4)^2$

Solution: Use Exponent's rules: $(x^a)^b = x^{a \times b}$
$(6x^2y^4)^2 = (6)^2(x^2)^2(y^4)^2 = 36x^{2 \times 2}y^{4 \times 2} = 36x^4y^8$

Example 2. Simplify. $\left(\dfrac{5x}{2x^2}\right)^2$

Solution: First, cancel the common factor: $x \rightarrow \left(\dfrac{5x}{2x^2}\right)^2 = \left(\dfrac{5}{2x}\right)^2$
Use Exponent's rules: $\left(\dfrac{a}{b}\right)^c = \dfrac{a^c}{b^c}$, Then: $\left(\dfrac{5}{2x}\right)^2 = \dfrac{5^2}{(2x)^2} = \dfrac{25}{4x^2}$

Example 3. Simplify. $\left(3x^5y^4\right)^2$

Solution: Use Exponent's rules: $(x^a)^b = x^{a \times b}$
$$\left(3x^5y^4\right)^2 = (3)^2\left(x^5\right)^2(y^4)^2 = 9x^{5 \times 2}y^{4 \times 2} = 9x^{10}y^8$$

Example 4. Simplify. $\left(\dfrac{2x}{3x^2}\right)^2$

Solution: First, cancel the common factor: $x \rightarrow \left(\dfrac{2x}{3x^2}\right)^2 = \left(\dfrac{2}{3x}\right)^2$
Use Exponent's rules: $\left(\dfrac{a}{b}\right)^c = \dfrac{a^c}{b^c}$, Then: $\left(\dfrac{2}{3x}\right)^2 = \dfrac{2^2}{(3x)^2} = \dfrac{4}{9x^2}$

ZERO AND NEGATIVE EXPONENTS

☑ Zero-Exponent Rule: $a^0 = 1$, this means that anything raised to the zero power is 1. For example: $(5xy)^0 = 1$

☑ A negative exponent simply means that the base is on the wrong side of the fraction line, so you need to flip the base to the other side. For instance, "x^{-2}" (pronounced as "ecks to the minus two") just means "x^2" but underneath, as in $\frac{1}{x^2}$.

Examples:

Example 1. Evaluate. $\left(\frac{2}{3}\right)^{-2} =$

Solution: Use negative exponent's rule: $\left(\frac{x^a}{x^b}\right)^{-2} = \left(\frac{x^b}{x^a}\right)^2 \rightarrow \left(\frac{2}{3}\right)^{-2} = \left(\frac{3}{2}\right)^2 =$
Then: $\left(\frac{3}{2}\right)^2 = \frac{3^2}{2^2} = \frac{9}{4}$

Example 2. Evaluate. $\left(\frac{4}{5}\right)^{-3} =$

Solution: Use negative exponent's rule: $\left(\frac{x^a}{x^b}\right)^{-2} = \left(\frac{x^b}{x^a}\right)^2 \rightarrow \left(\frac{4}{5}\right)^{-3} = \left(\frac{5}{4}\right)^3 =$
Then: $\left(\frac{5}{4}\right)^3 = \frac{5^3}{4^3} = \frac{125}{64}$

Example 3. Evaluate. $\left(\frac{x}{y}\right)^0 =$

Solution: Use zero-exponent Rule: $a^0 = 1$
Then: $\left(\frac{x}{y}\right)^0 = 1$

Example 4. Evaluate. $\left(\frac{5}{6}\right)^{-1} =$

Solution: Use negative exponent's rule: $\left(\frac{x^a}{x^b}\right)^{-2} = \left(\frac{x^b}{x^a}\right)^2 \rightarrow \left(\frac{5}{6}\right)^{-1} = \left(\frac{6}{5}\right)^1 = \frac{6}{5}$

NEGATIVE EXPONENTS AND NEGATIVE BASES

☑ A negative exponent is the reciprocal of that number with a positive exponent. $(3)^{-2} = \frac{1}{3^2}$

☑ To simplify a negative exponent, make the power positive!

☑ The parenthesis is important! -5^{-2} is not the same as $(-5)^{-2}$

$$-5^{-2} = -\frac{1}{5^2} \text{ and } (-5)^{-2} = +\frac{1}{5^2}$$

Examples:

Example 1. Simplify. $\left(\frac{5a}{6c}\right)^{-2} =$

Solution: Use negative exponent's rule: $\left(\frac{x^a}{x^b}\right)^{-2} = \left(\frac{x^b}{x^a}\right)^{2} \rightarrow \left(\frac{5a}{6c}\right)^{-2} = \left(\frac{6c}{5a}\right)^{2}$

Now use exponent's rule: $\left(\frac{a}{b}\right)^{c} = \frac{a^c}{b^c} \rightarrow = \left(\frac{6c}{5a}\right)^{2} = \frac{6^2c^2}{5^2a^2}$

Then: $\frac{6^2c^2}{5^2a^2} = \frac{36c^2}{25a^2}$

Example 2. Simplify. $\left(\frac{2x}{3yz}\right)^{-3} =$

Solution: Use negative exponent's rule: $\left(\frac{x^a}{x^b}\right)^{-2} = \left(\frac{x^b}{x^a}\right)^{2} \rightarrow \left(\frac{2x}{3yz}\right)^{-3} = \left(\frac{3yz}{2x}\right)^{3}$

Now use exponent's rule: $\left(\frac{a}{b}\right)^{c} = \frac{a^c}{b^c} \rightarrow \left(\frac{3yz}{2x}\right)^{3} = \frac{3^3y^3z^3}{2^3x^3} = \frac{27y^3z^3}{8x^3}$

Example 3. Simplify. $\left(\frac{3a}{2c}\right)^{-2} =$

Solution: Use negative exponent's rule: $\left(\frac{x^a}{x^b}\right)^{-2} = \left(\frac{x^b}{x^a}\right)^{2} \rightarrow \left(\frac{3a}{2c}\right)^{-2} = \left(\frac{2c}{3a}\right)^{2}$

Now use exponent's rule: $\left(\frac{a}{b}\right)^{c} = \frac{a^c}{b^c} \rightarrow = \left(\frac{2c}{3a}\right)^{2} = \frac{2^2c^2}{3^2a^2}$

Then: $\frac{2^2c^2}{3^2a^2} = \frac{4c^2}{9a^2}$

SCIENTIFIC NOTATION

☑ Scientific notation is used to write very big or very small numbers in decimal form.

☑ In scientific notation, all numbers are written in the form of: $m \times 10^n$, where m is greater than 1 and less than 10.

☑ To convert a number from scientific notation to standard form, move the decimal point to the left (if the exponent of ten is a negative number), or to the right (if the exponent is positive).

Examples:

Example 1. Write 0.00015 in scientific notation.

Solution: First, move the decimal point to the right so you have a number between 1 and 10. That number is 1.5. Now, determine how many places the decimal moved in step 1 by the power of 10. We moved the decimal point 4 digits to the right. Then: $10^{-4} \rightarrow$ When the decimal moved to the right, the exponent is negative. Then: $0.00015 = 1.5 \times 10^{-4}$

Example 2. Write 9.5×10^{-5} in standard notation.

Solution: $10^{-5} \rightarrow$ When the decimal moved to the right, the exponent is negative. Then: $9.5 \times 10^{-5} = 0.000095$

Example 3. Write 0.00012 in scientific notation.

Solution: First, move the decimal point to the right so you have a number between 1 and 10. Then: $m = 1.2$, Now, determine how many places the decimal moved in step 1 by the power of 10.
$10^{-4} \rightarrow$ Then: $0.00012 = 1.2 \times 10^{-4}$

Example 4. Write 8.3×10^5 in standard notation.
Solution: $10^{-5} \rightarrow$ The exponent is positive 5. Then, move the decimal point to the right five digits. (remember 8.3 = 8.30000),
Then: $8.3 \times 10^5 = 830000$

RADICALS

☑ If n is a positive integer and x is a real number, then: $\sqrt[n]{x} = x^{\frac{1}{n}}$,

$$\sqrt[n]{xy} = x^{\frac{1}{n}} \times y^{\frac{1}{n}}, \sqrt[n]{\frac{x}{y}} = \frac{x^{\frac{1}{n}}}{y^{\frac{1}{n}}}, \text{ and } \sqrt[n]{x} \times \sqrt[n]{y} = \sqrt[n]{xy}$$

☑ A square root of x is a number r whose square is: $r^2 = x$ (r is a square root of x)

☑ To add and subtract radicals, we need to have the same values under the radical. For example: $\sqrt{3} + \sqrt{3} = 2\sqrt{3}$, $3\sqrt{5} - \sqrt{5} = 2\sqrt{5}$

Examples:

Example 1. Find the square root of $\sqrt{169}$.

Solution: First, factor the number: $169 = 13^2$, Then: $\sqrt{169} = \sqrt{13^2}$,
Now use radical rule: $\sqrt[n]{a^n} = a$. Then: $\sqrt{169} = \sqrt{13^2} = 13$

Example 2. Evaluate. $\sqrt{9} \times \sqrt{25} =$

Solution: Find the values of $\sqrt{9}$ and $\sqrt{25}$. Then: $\sqrt{9} \times \sqrt{25} = 3 \times 5 = 15$

Example 3. Solve. $7\sqrt{2} + 4\sqrt{2}$.

Solution: Since we have the same values under the radical, we can add these two radicals: $7\sqrt{2} + 4\sqrt{2} = 11\sqrt{2}$

Example 4. Evaluate. $\sqrt{2} \times \sqrt{8} =$

Solution: Use this radical rule: $\sqrt[n]{x} \times \sqrt[n]{y} = \sqrt[n]{xy} \rightarrow \sqrt{2} \times \sqrt{8} = \sqrt{16}$
The square root of 16 is 4. Then: $\sqrt{2} \times \sqrt{8} = \sqrt{16} = 4$

CHAPTER 9: PRACTICES

✍ Find the products.

1) $2x^3 \times 4xy^2 =$

2) $6x^2y \times 8x^2y^2 =$

3) $5x^3y^2 \times 3x^2y^3 =$

4) $7xy^4 \times 4x^2y =$

5) $3x^4y^5 \times 9x^3y^2 =$

6) $6x^3y^2 \times 7x^3y^3 =$

7) $4x^3y^6 \times 2x^4y^2 =$

8) $7x^4y^3 \times 3x^3y^2 =$

9) $10x^5y^2 \times 10x^4y^3 =$

10) $8x^2y^3 \times 5x^6y^2 =$

11) $9y^5 \times 2x^6y^3 =$

12) $7x^4 \times 7x^2y^2 =$

✍ Simplify.

13) $\frac{3^3 \times 3^4}{3^9 \times 3} =$

14) $\frac{6x}{30x^2} =$

15) $\frac{18x^4}{6x^3} =$

16) $\frac{42x^3}{56x^3y^2} =$

17) $\frac{18y^3}{54x^4y^4} =$

18) $\frac{150x^3y^5}{50x^2y^3} =$

19) $\frac{2^3 \times 2^2}{7^2 \times 7} =$

20) $\frac{12x}{2x^2} =$

21) $\frac{25x^6}{5x^3} =$

22) $\frac{48y^4}{56x^5y^3} =$

✍ Solve.

23) $(3x^2y^6)^3 =$

24) $(2x^3y^4)^5 =$

25) $(2x \times 5xy^2)^2 =$

26) $(3x \times 2y^3)^2 =$

27) $\left(\frac{8x}{x^3}\right)^3 =$

28) $\left(\frac{9y}{3y^2}\right)^3 =$

29) $\left(\frac{6x^3y^4}{2x^4y^2}\right)^3 =$

30) $\left(\frac{27\ ^4y^4}{54\ ^3y^5}\right)^2 =$

31) $\left(\frac{9x^8y^4}{3x^5y^2}\right)^2 =$

32) $\left(\frac{35\ ^7y^4}{7x^5y^3}\right)^2 =$

✏ Evaluate each expression. (Zero and Negative Exponents)

33) $\left(\frac{1}{8}\right)^{-3} =$

34) $\left(\frac{1}{6}\right)^{-2} =$

35) $\left(\frac{3}{4}\right)^{-2} =$

36) $\left(\frac{4}{9}\right)^{-2} =$

37) $\left(\frac{1}{4}\right)^{-4} =$

38) $\left(\frac{2}{7}\right)^{-3} =$

✏ Write each expression with positive exponents.

39) $18x^{-2}y^{-6} =$

40) $35x^{-3}y^{-5} =$

41) $-12y^{-4} =$

42) $-25x^{-6} =$

43) $15a^{-3}b^6 =$

44) $20a^6b^{-5}c^{-3} =$

45) $46x^6y^{-3}z^{-7} =$

46) $\frac{16y}{x^3y^{-3}} =$

47) $\frac{24a^{-3}b}{-16c^{-3}} =$

✏ Write each number in scientific notation.

48) $0.00521 =$

49) $0.000067 =$

50) $25,000 =$

51) $36,000,000 =$

✏ Evaluate.

52) $\sqrt{6} \times \sqrt{6} =$ ....

53) $\sqrt{49} - \sqrt{4} =$ ....

54) $\sqrt{36} + \sqrt{64} =$ ....

55) $\sqrt{9} \times \sqrt{49} =$ ....

56) $\sqrt{2} \times \sqrt{18} =$ ....

57) $3\sqrt{5} + 2\sqrt{5} =$ ....

CHAPTER 9: ANSWERS

1) $8x^4y^2$

2) $48x^4y^3$

3) $15x^5y^5$

4) $28x^3y^5$

5) $27x^7y^7$

6) $42x^6y^5$

7) $8x^7y^8$

8) $21x^7y^5$

9) $100x^9y^5$

10) $40x^8y^5$

11) $18x^6y^8$

12) $49x^6y^2$

13) $\frac{1}{27}$

14) $\frac{1}{5x}$

15) $3x$

16) $\frac{3}{4y^2}$

17) $\frac{1}{3x^4y}$

18) $3xy^2$

19) $\frac{32}{343}$

20) $\frac{6}{x}$

21) $5x^3$

22) $\frac{6y}{7x^5}$

23) $27x^6y^{18}$

24) $32x^{15}y^{20}$

25) $100x^4y^4$

26) $36x^2y^6$

27) $\frac{512}{x^6}$

28) $\frac{27}{y^3}$

29) $\frac{27^{6}}{x^3}$

30) $\frac{x^2}{4y^2}$

31) $9x^6y^4$

32) $25x^4y^2$

33) 512

34) 36

35) $\frac{16}{9}$

36) $\frac{81}{16}$

37) 256

38) $\frac{343}{8}$

39) $\frac{18}{x^2y^6}$

40) $\frac{35}{x^3y^5}$

41) $-\frac{12}{y^4}$

42) $-\frac{25}{x^6}$

43) $\frac{15b^6}{a^3}$

44) $\frac{20^{6}}{b^5c^3}$

45) $\frac{46x^6}{y^3z^7}$

46) $\frac{16y^4}{x^3}$

47) $-\frac{3bc^3}{2a^3}$

48) 5.21×10^{-3}

49) 6.7×10^{-5}

50) 25×10^3

51) 36×10^6

52) 6

53) 5

54) 14

55) 21

56) 6

57) $5\sqrt{5}$

CHAPTER 10:

POLYNOMIALS

Math Topics that you'll learn in this chapter:

▶ Simplifying Polynomials

▶ Adding and Subtracting Polynomials

▶ Multiplying Monomials

▶ Multiplying and Dividing Monomials

▶ Multiplying a Polynomial and a Monomial

▶ Multiplying Binomials

▶ Factoring Trinomials

Simplifying Polynomials

☑ To simplify Polynomials, find "like" terms. (they have same variables with same power).

☑ Use "FOIL". (First–Out–In–Last) for binomials:

$$(x + a)(x + b) = x^2 + (b + a)x + ab$$

☑ Add or Subtract "like" terms using order of operation.

Examples:

Example 1. Simplify this expression. $x(2x + 5) + 6x =$

Solution: Use Distributive Property: $x(2x + 5) = 2x^2 + 5x$

Now, combine like terms: $x(2x + 5) + 6x = 2x^2 + 5x + 6x = 2x^2 + 11x$

Example 2. Simplify this expression. $(x + 2)(x + 3) =$

Solution: First, apply the FOIL method: $(a + b)(c + d) = ac + ad + bc + bd$

$(x + 2)(x + 3) = x^2 + 3x + 2x + 6$

Now combine like terms: $x^2 + 3x + 2x + 6 = x^2 + 5x + 6$

Example 3. Simplify this expression. $4x(2x - 3) + 6x^2 - 4x =$

Solution: Use Distributive Property: $4x(2x - 3) = 8x^2 - 12x$

Then: $4x(2x - 3) + 6x^2 - 4x = 8x^2 - 12x + 6x^2 - 4x$

Now combine like terms: $8x^2 + 6x^2 = 14x^2$, and $-12x - 4x = -16x$

The simplified form of the expression: $8x^2 - 12x + 6x^2 - 4x = 14x^2 - 16x$

ADDING AND SUBTRACTING POLYNOMIALS

☑ Adding polynomials is just a matter of combining like terms, with some order of operations considerations thrown in.

☑ Be careful with the minus signs, and don't confuse addition and multiplication!

☑ For subtracting polynomials, sometimes you need to use the Distributive Property: $a(b + c) = ab + ac$, $a(b - c) = ab - ac$

Examples:

Example 1. Simplify the expressions. $(x^3 - 3x^4) - (2x^4 - 5x^3) =$

Solution: First, use Distributive Property:
$-(2x^4 - 5x^3) = -1(2x^4 - 5x^3) = -2x^4 + 5x^3$
$\rightarrow (x^3 - 3x^4) - (2x^4 - 5x^3) = x^3 - 3x^4 - 2x^4 + 5x^3$
Now combine like terms: $x^3 + 5x^3 = 6x^3$ and $-3x^4 - 2x^4 = -5x^4$
Then: $(x^3 - 3x^4) - (2x^4 - 5x^3) = x^3 - 3x^4 - 2x^4 + 5x^3 = 6x^3 - 5x^4$
Write the answer in standard form: $6x^3 - 5x^4 = -5x^4 + 6x^3$

Example 2. Add expressions. $(2x^3 - 4) + (6x^3 - 2x^2) =$

Solution: Remove parentheses:
$(2x^3 - 4) + (6x^3 - 2x^2) = 2x^3 - 4 + 6x^3 - 2x^2$
Now combine like terms: $2x^3 - 4 + 6x^3 - 2x^2 = 8x^3 - 2x^2 - 4$

Example 3. Simplify the expressions. $(8x^2 - 3x^3) - (2x^2 + 5x^3) =$

Solution: First, use Distributive Property:
$-(2x^2 + 5x^3) = -2x^2 - 5x^3 \rightarrow (8x^2 - 3x^3) - (2x^2 + 5x^3) = 8x^2 - 3x^3 - 2x^2 - 5x^3$
Now combine like terms and write in standard form:
$8x^2 - 3x^3 - 2x^2 - 5x^3 = -8x^3 + 6x^2$

MULTIPLYING MONOMIALS

☑ A monomial is a polynomial with just one term: Examples: $2x$ or $7y^2$.

☑ When you multiply monomials, first multiply the coefficients (a number placed before and multiplying the variable) and then multiply the variables using multiplication property of exponents.

$$x^a \times x^b = x^{a+b}$$

Examples:

Example 1. Multiply expressions. $5xy^4z^2 \times 3x^2y^5z^3$

Solution: Find the same variables and use multiplication property of exponents: $x^a \times x^b = x^{a+b}$
$x \times x^2 = x^{1+2} = x^3$, $y^4 \times y^5 = y^{4+5} = y^9$ and $z^2 \times z^3 = z^{2+3} = z^5$
Then, multiply coefficients and variables: $5xy^4z^2 \times 3x^2y^5z^3 = 15x^3y^9z^5$

Example 2. Multiply expressions. $-2a^5b^4 \times 8a^3b^4 =$

Solution: Use the multiplication property of exponents: $x^a \times x^b = x^{a+b}$
$a^5 \times a^3 = a^{5+3} = a^8$ and $b^4 \times b^4 = b^{4+4} = b^8$
Then: $-2a^5b^4 \times 8a^3b^4 = -16a^8b^8$

Example 3. Multiply. $7xy^3z^5 \times 4x^2y^4z^3$

Solution: Use the multiplication property of exponents: $x^a \times x^b = x^{a+b}$
$x \times x^2 = x^{1+2} = x^3$, $y^3 \times y^4 = y^{3+4} = y^7$ and $z^5 \times z^3 = z^{5+3} = z^8$
Then: $7xy^3z^5 \times 4x^2y^5z^3 = 28x^3y^7z^8$

Example 4. Simplify. $(5a^6b^3)(-9a^7b^2) =$

Solution: Use the multiplication property of exponents: $x^a \times x^b = x^{a+b}$
$a^6 \times a^7 = a^{6+7} = a^{13}$ and $b^3 \times b^2 = b^{3+2} = b^5$
Then: $(5a^6b^3) \times (-9a^6b^2) = -45a^{13}b^5$

MULTIPLYING AND DIVIDING MONOMIALS

✅ When you divide or multiply two monomials, you need to divide or multiply their coefficients and then divide or multiply their variables.

✅ In case of exponents with the same base, for Division, subtract their powers, for Multiplication, add their powers.

✅ Exponent's Multiplication and Division rules:

$$x^a \times x^b = x^{a+b} , \qquad \frac{x^a}{x^b} = x^{a-b}$$

Examples:

Example 1. Multiply expressions. $(-5x^8)(4x^6) =$

Solution: Use multiplication property of exponents:
$x^a \times x^b = x^{a+b} \rightarrow x^8 \times x^6 = x^{14}$
Then: $(-5x^5)(4x^4) = -20x^{14}$

Example 2. Divide expressions. $\frac{14x^5y^4}{2xy^3} =$

Solution: Use division property of exponents:
$\frac{x^a}{x^b} = x^{a-b} \rightarrow \frac{x^5}{x} = x^{5-1} = x^4$ and $\frac{y^4}{y^3} = y$
Then: $\frac{14x^5y^4}{2xy^3} = 7x^4y$

Example 3. Divide expressions. $\frac{56a^8b^3}{8ab^3}$

Solution: Use division property of exponents:
$\frac{x^a}{x^b} = x^{a-b} \rightarrow \frac{a^8}{a} = a^{8-1} = a^7$ and $\frac{b^3}{b^3} = 1$
Then: $\frac{56a^8b^3}{8ab^3} = 7a^7$

Multiplying a Polynomial and a Monomial

☑ When multiplying monomials, use the product rule for exponents.

$$x^a \times x^b = x^{a+b}$$

☑ When multiplying a monomial by a polynomial, use the distributive property.

$$a \times (b + c) = a \times b + a \times c = ab + ac$$
$$a \times (b - c) = a \times b - a \times c = ab - ac$$

Examples:

Example 1. Multiply expressions. $5x(3x - 2)$

Solution: Use Distributive Property:
$$5x(3x - 2) = 5x \times 3x - 5x \times (-2) = 15x^2 - 10x$$

Example 2. Multiply expressions. $x(2x^2 + 3y^2)$

Solution: Use Distributive Property:
$$x(2x^2 + 3y^2) = x \times 2x^2 + x \times 3y^2 = 2x^3 + 3xy^2$$

Example 3. Multiply. $-4x(-5x^2 + 3x - 6)$

Solution: Use Distributive Property:
$$-4x(-5x^2 + 3x - 6) = (-4x)(-5x^2) + (-4x) \times (3x) + (-4x) \times (-6) =$$
Now simplify:
$$(-4x)(-5x^2) + (-4x) \times (3x) + (-4x) \times (-6) = 20x^3 - 12x^2 + 24x$$

MULTIPLYING BINOMIALS

☑ A binomial is a polynomial that is the sum or the difference of two terms, each of which is a monomial.

☑ To multiply two binomials, use the "FOIL" method. (First–Out–In–Last)

$$(x + a)(x + b) = x \times x + x \times b + a \times x + a \times b = x^2 + bx + ax + ab$$

Examples:

Example 1. Multiply Binomials. $(x + 2)(x - 4) =$

Solution: Use "FOIL". (First–Out–In–Last):
$(x + 2)(x - 4) = x^2 - 4x + 2x - 8$
Then combine like terms: $x^2 - 4x + 2x - 8 = x^2 - 2x - 8$

Example 2. Multiply. $(x - 5)(x - 2) =$

Solution: Use "FOIL". (First–Out–In–Last):
$(x - 5)(x - 2) = x^2 - 2x - 5x + 10$
Then simplify: $x^2 - 2x - 5x + 10 = x^2 - 7x + 10$

Example 3. Multiply. $(x - 3)(x + 6) =$

Solution: Use "FOIL". (First–Out–In–Last):
$(x - 3)(x + 6) = x^2 + 6x - 3x - 18$
Then simplify: $x^2 + 6x - 3x - 18 = x^2 + 3x - 18$

Example 4. Multiply Binomials. $(x + 8)(x + 4) =$

Solution: Use "FOIL". (First–Out–In–Last):
$(x + 8)(x + 4) = x^2 + 4x + 8x + 32$
Then combine like terms: $x^2 + 4x + 8x + 32 = x^2 + 12x + 32$

FACTORING TRINOMIALS

To factor trinomials, you can use following methods:

☑ "FOIL": $(x + a)(x + b) = x^2 + (b + a)x + ab$

☑ "Difference of Squares":

$$a^2 - b^2 = (a + b)(a - b)$$
$$a^2 + 2ab + b^2 = (a + b)(a + b)$$
$$a^2 - 2ab + b^2 = (a - b)(a - b)$$

☑ "Reverse FOIL": $x^2 + (b + a)x + ab = (x + a)(x + b)$

Examples:

Example 1. Factor this trinomial. $x^2 - 2x - 8$

Solution: Break the expression into groups. You need to find two numbers that their product is -8 and their sum is -2. (remember "Reverse FOIL": $x^2 + (b + a)x + ab = (x + a)(x + b)$). Those two numbers are 2 and -4. Then: $x^2 - 2x - 8 = (x^2 + 2x) + (-4x - 8)$
Now factor out x from $x^2 + 2x$: $x(x + 2)$, and factor out -4 from $-4x - 8$: $-4(x + 2)$; Then: $(x^2 + 2x) + (-4x - 8) = x(x + 2) - 4(x + 2)$
Now factor out like term: $(x + 2)$. Then: $(x + 2)(x - 4)$

Example 2. Factor this trinomial. $X^2 - 2x - 24$

Solution: Break the expression into groups: $(x^2 + 4x) + (-6x - 24)$
Now factor out x from $x^2 + 4x$: $x(x + 4)$, and factor out -6 from $-6x - 24$: $-6(x + 4)$; Then: $(x + 4) - 6(x + 4)$, now factor out like term: $(x = 4) \rightarrow x(x + 4) - 6(x + 4) = (x + 4)(x - 6)$

CHAPTER 10: PRACTICES

✎ Simplify each polynomial.

1) $2(5x + 7) =$

2) $6(3x - 9) =$

3) $x(6x + 3) + 4x =$

4) $2x(x + 8) + 6x =$

5) $8x(2x + 1) - 6x =$

6) $5x(4x - 2) + 2x^2 - 1 =$

7) $4x^2 - 6 - 8x(2x + 7) =$

8) $7x^2 + 9 - 3x(x + 4) =$

✎ Add or subtract polynomials.

9) $(5x^2 + 4) + (3x^2 - 6) =$

10) $(2x^2 - 7x) - (4x^2 + 3x) =$

11) $(8x^3 - 5x^2) + (2x^3 - 6x^2) =$

12) $(3x^3 - 6x) - (7x^3 - 2x) =$

13) $(15x^3 + 3x^2) + (12x^2 - 9) =$

14) $(5x^3 - 8) - (2x^3 - 9x^2) =$

15) $(6x^3 + 2x) + (3x^3 - 2x) =$

16) $(3x^3 - 7x) - (4x^3 + 6x) =$

✎ Find the products. (Multiplying Monomials)

17) $6x^2 \times 4x^3 =$

18) $5x^4 \times 6x^3 =$

19) $-5a^4b \times 4ab^3 =$

20) $(-6x^3yz) \times (-5xy^2z^4) =$

21) $-a^5bc \times a^2b^4 =$

22) $7u^3t^2 \times (-8ut) =$

23) $10x^2z \times 4xy^3 =$

24) $12x^3z \times 2xy^5 =$

25) $-4a^3bc \times a^4b^3 =$

26) $8x^6y^2 \times (-10xy) =$

✍ **Simplify each expression. (Multiplying and Dividing Monomials)**

27) $(6x^2y^3)(9x^4y^2) =$

28) $(3x^3y^2)(7x^4y^3) =$

29) $(12x^8y^5)(4x^5y^7) =$

30) $(10a^3b^2)(5a^3b^8) =$

31) $\frac{32x^4y^2}{8x^3y} =$

32) $\frac{48x^5y^6}{6x^2y} =$

33) $\frac{72x^{15}y^{10}}{9x^8y^6} =$

34) $\frac{200\ ^8y^{12}}{5x^4y^8} =$

✍ **Find each product. (Multiplying a Polynomial and a Monomial)**

35) $2x(4x - y) =$

36) $6x(2x + 5y) =$

37) $6x(x - 9y) =$

38) $x(4x^2 + 3x - 8) =$

39) $6x(-2x^2 + 6x + 3) =$

40) $9x(3x^2 - 6x - 10) =$

✍ **Find each product. (Multiplying Binomials)**

41) $(x - 4)(x + 4) =$

42) $(x - 6)(x - 5) =$

43) $(x + 8)(x + 2) =$

44) $(x - 8)(x + 9) =$

45) $(x + 4)(x - 6) =$

46) $(x - 12)(x + 4) =$

✍ **Factor each trinomial.**

47) $x^2 + x - 12 =$

48) $x^2 + 3x - 10 =$

49) $x^2 - 10x - 24 =$

50) $x^2 + 19x + 48 =$

51) $2x^2 - 14x + 24 =$

52) $3x^2 + 3x - 18 =$

CHAPTER 10: ANSWERS

1) $10x + 14$

2) $18x - 54$

3) $6x^2 + 7x$

4) $2x^2 + 22x$

5) $16x^2 + 2x$

6) $22x^2 - 10x - 1$

7) $-12x^2 - 56x - 6$

8) $4x^2 - 12x + 9$

9) $8x^2 - 2$

10) $-2x^2 - 10x$

11) $10x^3 - 11x^2$

12) $-4x^2 - 4x$

13) $15x^3 + 15x^2 - 9$

14) $3x^3 + 9x^2 - 8$

15) $9x^3$

16) $-x^3 - 13x^2$

17) $24x^5$

18) $30x^7$

19) $-20a^5b^4$

20) $30x^4y^3z^5$

21) $-a^7b^5c$

22) $-56u^4t^3$

23) $40x^3y^3z$

24) $24x^4y^5z$

25) $-4a^7b^4c$

26) $-80x^7y^3$

27) $54x^6y^5$

28) $21x^7y^5$

29) $48x^{13}y^{12}$

30) $50a^6b^{10}$

31) $4xy$

32) $8x^3y^5$

33) $8x^7y^4$

34) $40x^4y^4$

35) $8x^2 - 2xy$

36) $12x^2 + 30xy$

37) $6x^2 - 54xy$

38) $4x^3 + 3x^2 - 8x$

39) $-12x^3 + 36x^2 + 18x$

40) $27x^3 - 54x^2 - 90x$

41) $x^2 - 16$

42) $x^2 - 11x + 30$

43) $x^2 + 10x + 16$

44) $x^2 + x - 72$

45) $x^2 - 2x - 24$

46) $x^2 - 8x - 48$

47) $(x + 4)(x - 3)$

48) $(x + 5)(x - 2)$

49) $(x - 12)(x + 2)$

50) $(x + 16)(x + 3)$

51) $(2x - 8)(x - 3)$

52) $(3x - 6)(x + 3)$

CHAPTER 11:

GEOMETRY AND SOLID FIGURES

Math Topics that you'll learn in this chapter:

▶ The Pythagorean Theorem

▶ Triangles

▶ Polygons

▶ Circles

▶ Trapezoids

▶ Cubes

▶ Rectangle Prisms

▶ Cylinder

The Pythagorean Theorem

☑ You can use the Pythagorean Theorem to find a missing side in a right triangle.

☑ In any right triangle: $a^2 + b^2 = c^2$

Examples:

Example 1. Right triangle ABC (not shown) has two legs of lengths 6 cm (AB) and 8 cm (AC). What is the length of the hypotenuse of the triangle (side BC)?

Solution: Use Pythagorean Theorem: $a^2 + b^2 = c^2$, $a = 6$, and $b = 8$
Then: $a^2 + b^2 = c^2 \rightarrow 6^2 + 8^2 = c^2 \rightarrow 36 + 64 = c^2 \rightarrow 100 = c^2 \rightarrow c = \sqrt{100} = 10$
The length of the hypotenuse is 10 cm.

Example 2. Find the hypotenuse of this triangle.

Solution: Use Pythagorean Theorem: $a^2 + b^2 = c^2$
Then: $a^2 + b^2 = c^2 \rightarrow 12^2 + 5^2 = c^2 \rightarrow 144 + 25 = c^2$
$c^2 = 169 \rightarrow c = \sqrt{169} = 13$

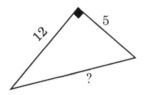

Example 3. Find the length of the missing side in this triangle.

Solution: Use Pythagorean Theorem: $a^2 + b^2 = c^2$
Then: $a^2 + b^2 = c^2 \rightarrow 3^2 + b^2 = 5^2 \rightarrow 9 + b^2 = 25 \rightarrow$
$b^2 = 25 - 9 \rightarrow b^2 = 16 \rightarrow b = \sqrt{16} = 4$

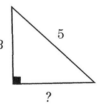

TRIANGLES

☑ In any triangle, the sum of all angles is 180 degrees.

☑ Area of a triangle = $\frac{1}{2}$ (*base* × *height*)

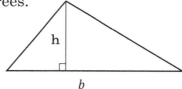

Examples:

What is the area of the following triangles?

Example 1.

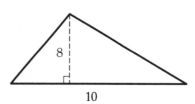

Solution: Use the area formula:
Area = $\frac{1}{2}$ (*base* × *height*)

base = 10 and *height* = 8
Area = $\frac{1}{2}(10 \times 8) = \frac{1}{2}(80) = 40$

Example 2.

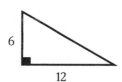

Solution: Use the area formula:
Area = $\frac{1}{2}$ (*base* × *height*)

base = 12 and *height* = 6 ; Area = $\frac{1}{2}(12 \times 6) = \frac{72}{2} = 36$

Example 3. What is the missing angle in this triangle?

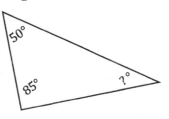

Solution:

In any triangle, the sum of all angles is 180 degrees. Let x be the missing angle.
Then: $50 + 85 + x = 180$;
$\rightarrow 135 + x = 180 \rightarrow x = 180 - 135 = 45$
The missing angle is 45 degrees.

POLYGONS

☑ The perimeter of a square = $4 \times side = 4s$

☑ The perimeter of a rectangle= $2(width + length)$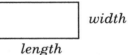

☑ The perimeter of trapezoid= $a + b + c + d$

☑ The perimeter of a regular hexagon = $6a$

☑ The perimeter of a parallelogram = $2(l + w)$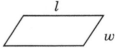

Examples:

Example 1. Find the perimeter of following regular hexagon.

Solution: Since the hexagon is regular, all sides are equal.

Then: The perimeter of The hexagon = $6 \times (one\ side)$

The perimeter of The hexagon = $6 \times (one\ side) = 6 \times 4 = 24\ m$

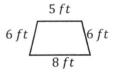

Example 2. Find the perimeter of following trapezoid.

Solution: The perimeter of a trapezoid = $a + b + c + d$

The perimeter of the trapezoid = $5 + 6 + 6 + 8 = 25\ ft$

CIRCLES

☑ In a circle, variable r is usually used for the radius and d for diameter.

☑ *Area of a circle* $= \pi r^2$ (π is about 3.14)

☑ *Circumference of a circle* $= 2\pi r$

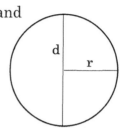

Examples:

Example 1. Find the area of this circle.

Solution:
Use area formula: $Area = \pi r^2$
$r = 8\ in \rightarrow Area = \pi(8)^2 = 64\pi$, $\pi = 3.14$
Then: $Area = 64 \times 3.14 = 200.96\ in^2$

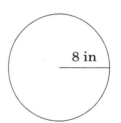

Example 2. Find the Circumference of this circle.

Solution:
Use Circumference formula: $Circumference = 2\pi r$
$r = 5\ cm \rightarrow Circumference = 2\pi(5) = 10\pi$
$\pi = 3.14$ Then: $Circumference = 10 \times 3.14 = 31.4\ cm$

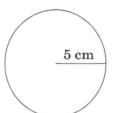

Example 3. Find the area of the circle.

Solution:
Use area formula: $Area = \pi r^2$,
$r = 5\ in$ then: $Area = \pi(5)^2 = 25\pi$, $\pi = 3.14$
Then: $Area = 25 \times 3.14 = 78.5$

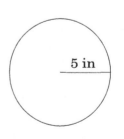

TRAPEZOIDS

☑ A quadrilateral with at least one pair of parallel sides is a trapezoid.

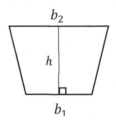

☑ Area of a trapezoid $= \frac{1}{2}h(b_1 + b_2)$

Examples:

Example 1. Calculate the area of this trapezoid.

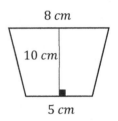

Solution:

Use area formula: A $= \frac{1}{2}h(b_1 + b_2)$

$b_1 = 5\ cm$, $b_2 = 8\ cm$ and $h = 10\ cm$

Then: A $= \frac{1}{2}(10)(8 + 5) = 5(13) = 65\ cm^2$

Example 2. Calculate the area of this trapezoid.

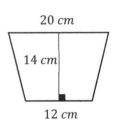

Solution:

Use area formula: A $= \frac{1}{2}h(b_1 + b_2)$

$b_1 = 12\ cm$, $b_2 = 20\ cm$ and $h = 14\ cm$

Then: A $= \frac{1}{2}(14)(12 + 20) = 7(32) = 224\ cm^2$

CUBES

☑ A cube is a three-dimensional solid object bounded by six square sides.

☑ Volume is the measure of the amount of space inside of a solid figure, like a cube, ball, cylinder or pyramid.

☑ The volume of a cube = $(one\ side)^3$

☑ The surface area of a cube = $6 \times (one\ side)^2$

Examples:

Example 1. Find the volume and surface area of this cube.

Solution: Use volume formula: $volume = (one\ side)^3$
Then: $volume = (one\ side)^3 = (2)^3 = 8\ cm^3$
Use surface area formula:
$surface\ area\ of\ cube: 6(one\ side)^2 = 6(2)^2 = 6(4) = 24\ cm^2$

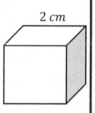

Example 2. Find the volume and surface area of this cube.

Solution: Use volume formula: $volume = (one\ side)^3$
Then: $volume = (one\ side)^3 = (5)^3 = 125\ cm^3$
Use surface area formula:
$surface\ area\ of\ cube: 6(one\ side)^2 = 6(5)^2 = 6(25) = 150\ cm^2$

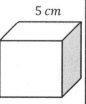

Example 3. Find the volume and surface area of this cube.

Solution: Use volume formula: $volume = (one\ side)^3$
Then: $volume = (one\ side)^3 = (7)^3 = 343\ m^3$
Use surface area formula:
$surface\ area\ of\ cube: 6(one\ side)^2 = 6(7)^2 = 6(49) = 294\ m^2$

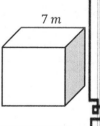

RECTANGULAR PRISMS

☑ A rectangular prism is a solid 3-dimensional object with six rectangular faces.

☑ The volume of a Rectangular prism = *Length × Width × Height*

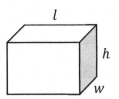

$Volume = l \times w \times h$

$Surface\ area = 2 \times (wh + lw + lh)$

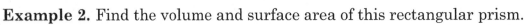

Examples:

Example 1. Find the volume and surface area of this rectangular prism.

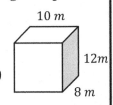

Solution: Use volume formula: $Volume = l \times w \times h$

Then: $Volume = 8 \times 6 \times 10 = 480\ m^3$

Use surface area formula: $Surface\ area = 2 \times (wh + lw + lh)$

Then: $Surface\ area = 2 \times \big((6 \times 10) + (8 \times 6) + (8 \times 10)\big)$

$= 2 \times (60 + 48 + 80) = 2 \times (188) = 376\ m^2$

Example 2. Find the volume and surface area of this rectangular prism.

Solution: Use volume formula: $Volume = l \times w \times h$

Then: $Volume = 10 \times 8 \times 12 = 960\ m^3$

Use surface area formula: $Surface\ area = 2 \times (wh + lw + lh)$

Then: $Surface\ area = 2 \times \big((8 \times 12) + (10 \times 8) + (10 \times 12)\big)$

$= 2 \times (96 + 80 + 120) = 2 \times (296) = 592\ m^2$

CYLINDER

☑ A cylinder is a solid geometric figure with straight parallel sides and a circular or oval cross-section.

☑ *Volume of a Cylinder* $= \pi(radius)^2 \times height$, $\pi \approx 3.14$

☑ *Surface area of a cylinder* $= 2\pi r^2 + 2\pi rh$

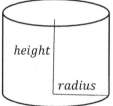

Examples:

Example 1. Find the volume and Surface area of the follow Cylinder.

Solution: Use volume formula:

$Volume = \pi(radius)^2 \times height$

Then: $Volume = \pi(3)^2 \times 8 = 9\pi \times 8 = 72\pi$

$\pi = 3.14$ then: $Volume = 72\pi = 72 \times 3.14 = 226.08 \ cm^3$

Use surface area formula: $Surface \ area = 2\pi r^2 + 2\pi rh$

Then: $2\pi(3)^2 + 2\pi(3)(8) = 2\pi(9) + 2\pi(24) = 18\pi + 48\pi = 66\pi$

$\pi = 3.14$ Then: $Surface \ area = 66 \times 3.14 = 207.24 \ cm^2$

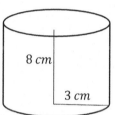

Example 2. Find the volume and Surface area of the follow Cylinder.

Solution: Use volume formula:

$Volume = \pi(radius)^2 \times height$

Then: $Volume = \pi(2)^2 \times 6 = \pi 4 \times 6 = 24\pi$

$\pi = 3.14$ then: $Volume = 24\pi = 75.36 \ cm^3$

Use surface area formula: $Surface \ area = 2\pi r^2 + 2\pi rh$

Then: $= 2\pi(2)^2 + 2\pi(2)(6) = 2\pi(4) + 2\pi(12) = 8\pi + 24\pi = 32\pi$

$\pi = 3.14$ then: $Surface \ area = 32 \times 3.14 = 100.48 \ cm^2$

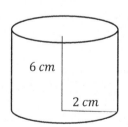

CHAPTER 11: PRACTICES

✎ Find the missing side?

1)
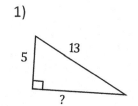

2)
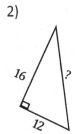

3)
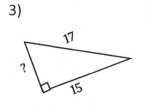

4)
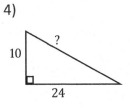

✎ Find the measure of the unknown angle in each triangle.

5)
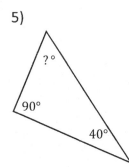

6)
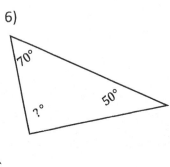

7)
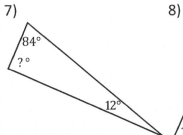

8)
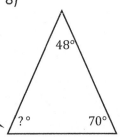

✎ Find the area of each triangle.

9)
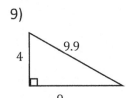

10)
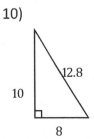

11)
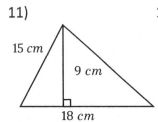

12)
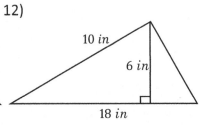

✎ Find the perimeter or circumference of each shape.

13)
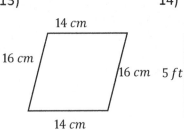

14)
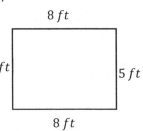

15)
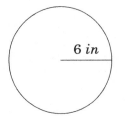

16) *regular hexagon*
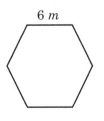

✎ Find the area of each trapezoid.

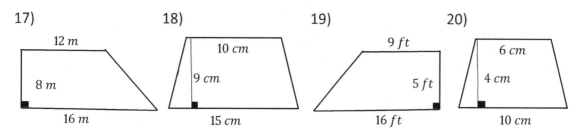

17)   12 m   8 m   16 m

18)   10 cm   9 cm   15 cm

19)   9 ft   5 ft   16 ft

20)   6 cm   4 cm   10 cm

✎ Find the volume of each cube.

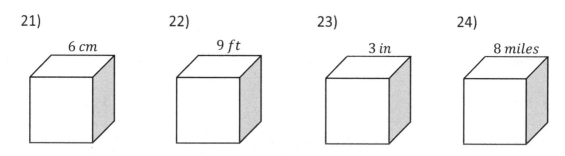

21)   6 cm

22)   9 ft

23)   3 in

24)   8 miles

✎ Find the volume of each Rectangular Prism.

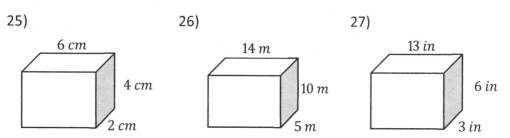

25)   6 cm   4 cm   2 cm

26)   14 m   10 m   5 m

27)   13 in   6 in   3 in

✎ Find the volume of each Cylinder. Round your answer to the nearest tenth. ($\pi = 3.14$)

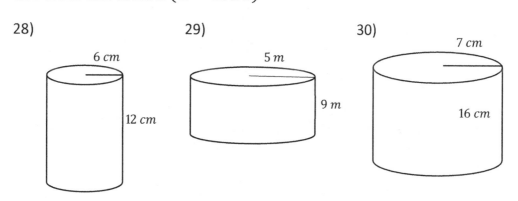

28)   6 cm   12 cm

29)   5 m   9 m

30)   7 cm   16 cm

CHAPTER 11: ANSWERS

1) 12

2) 20

3) 8

4) 26

5) 50

6) 60

7) 84

8) 62

9) 18

10) 40

11) $81\ cm^2$

12) $54in^2$

13) $60\ cm$

14) $26\ ft$

15) $12\ \pi \approx 37.68\ in$

16) $36\ m$

17) $112\ m^2$

18) $112.5\ cm^2$

19) $62.5\ ft^2$

20) $32\ cm^2$

21) $216\ cm^3$

22) $729\ ft^3$

23) $27\ in^3$

24) $512\ mi^3$

25) $48\ cm^3$

26) $700\ m^3$

27) $234\ in^3$

28) $1,356.5\ cm^3$

29) $706.5\ m^3$

30) $2,461.8\ cm^3$

CHAPTER 12:

STATISTICS

Math Topics that you'll learn in this chapter:

▶ Mean, Median, Mode, and Range of the Given Data

▶ Pie Graph

▶ Probability Problems

▶ Permutations and Combinations

MEAN, MEDIAN, MODE, AND RANGE OF THE GIVEN DATA

☑ Mean: $\dfrac{sum\ of\ the\ data}{total\ number\ of\ data\ entires}$

☑ Mode: the value in the list that appears most often

☑ Median: is the middle number of a group of numbers arranged in order by size.

☑ Range: the difference of the largest value and smallest value in the list

---

Examples:

Example 1. What is the mode of these numbers? $4, 5, 7, 5, 7, 4, 0, 4$

Solution: Mode: the value in the list that appears most often.
Therefore, the mode is number 4. There are three number 4 in the data.

Example 2. What is the median of these numbers? $5, 10, 14, 9, 16, 19, 6$

Solution: Write the numbers in order: $5, 6, 9, 10, 14, 16, 19$
The median is the number in the middle. Therefore, the median is 10.

Example 3. What is the mean of these numbers? $8, 2, 8, 5, 3, 2, 4, 8$

Solution: Mean: $\dfrac{sum\ of\ the\ data}{total\ number\ of\ data\ entires} = \dfrac{8+2+8+5+3+2+4+8}{8} = 5$

Example 4. What is the range in this list? $4, 9, 13, 8, 15, 18, 5$

Solution: Range is the difference of the largest value and smallest value in the list. The largest value is 18 and the smallest value is 4.
Then: $18 - 4 = 14$

PIE GRAPH

☑ A Pie Chart is a circle chart divided into sectors, each sector represents the relative size of each value.

☑ Pie charts represent a snapshot of how a group is broken down into smaller pieces.

Example:

A library has 820 books that include Mathematics, Physics, Chemistry, English and History. Use the following graph to answer the questions.

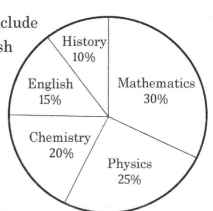

Example 1. What is the number of Mathematics books?

Solution: Number of total books = 820

Percent of Mathematics books = 30% = 0.30

Then, the number of Mathematics books: 0.30 × 820 = 246

Example 2. What is the number of History books?

Solution: Number of total books = 820

Percent of History books = 10% = 0.10

Then: 0.10 × 820 = 82

Example 3. What is the number of Chemistry books?

Solution: Number of total books = 820

Percent of Chemistry books = 20% = 0.20

Then: 0.20 × 820 = 164

PROBABILITY PROBLEMS

☑ Probability is the likelihood of something happening in the future. It is expressed as a number between zero (can never happen) to 1 (will always happen).

☑ Probability can be expressed as a fraction, a decimal, or a percent.

☑ Probability formula: $Probability = \dfrac{number\ of\ desired\ outcomes}{number\ of\ total\ outcomes}$

Examples:

Example 1. Anita's trick–or–treat bag contains 12 pieces of chocolate, 18 suckers, 18 pieces of gum, 24 pieces of licorice. If she randomly pulls a piece of candy from her bag, what is the probability of her pulling out a piece of sucker?

Solution: Probability $= \dfrac{number\ of\ desired\ outcomes}{number\ of\ total\ outcomes}$

Probability of pulling out a piece of sucker $= \dfrac{18}{12+18+18+24} = \dfrac{18}{72} = \dfrac{1}{4}$

Example 2. A bag contains 20 balls: four green, five black, eight blue, a brown, a red and one white. If 19 balls are removed from the bag at random, what is the probability that a brown ball has been removed?

Solution: If 19 balls are removed from the bag at random, there will be one ball in the bag. The probability of choosing a brown ball is 1 out of 20. Therefore, the probability of not choosing a brown ball is 19 out of 20 and the probability of having not a brown ball after removing 19 balls is the same.

PERMUTATIONS AND COMBINATIONS

☑ Factorials are products, indicated by an exclamation mark. For example, $4! = 4 \times 3 \times 2 \times 1$ (Remember that $0!$ is defined to be equal to 1.)

☑ Permutations: The number of ways to choose a sample of k elements from a set of n distinct objects where order does matter, and replacements are not allowed. For a permutation problem, use this formula:

$$_nP_k = \frac{n!}{(n-k)!}$$

☑ Combination: The number of ways to choose a sample of r elements from a set of n distinct objects where order does not matter, and replacements are not allowed. For a combination problem, use this formula:

$$_nC_r = \frac{n!}{r!\,(n-r)!}$$

Examples:

Example 1. How many ways can the first and second place be awarded to 8 people?

Solution: Since the order matters, (the first and second place are different!) we need to use permutation formula where n is 8 and k is 2. Then: $\frac{n!}{(n-k)!} = \frac{8!}{(8-2)!} = \frac{8!}{6!} = \frac{8 \times 7 \times 6!}{6!}$, remove $6!$ from both sides of the fraction. Then: $\frac{8 \times 7 \times 6!}{6!} = 8 \times 7 = 56$

Example 2. How many ways can we pick a team of 2 people from a group of 6?

Solution: Since the order doesn't matter, we need to use a combination formula where n is 6 and r is 2. Then: $\frac{n!}{r!\,(n-r)!} = \frac{6!}{2!\,(6-2)!} = \frac{6!}{2!\,(4)!} = \frac{6 \times 5 \times 4!}{2!\,(4)!} = \frac{6 \times 5}{2 \times 1} = \frac{30}{2} = 15$

CHAPTER 12: PRACTICES

✍ Find the values of the Given Data.

1) 7, 10, 4, 2, 7

   Mode: _____   Range: _____

   Mean: _____   Median: _____

2) 4, 8, 2, 9, 8, 5

   Mode: _____   Range: _____

   Mean: _____   Median: _____

3) 12, 2, 6, 10, 6, 15

   Mode: _____   Range: _____

   Mean: _____   Median: _____

4) 12, 5, 1, 10, 2, 11, 1

   Mode: _____   Range: _____

   Mean: _____   Median: _____

✍ The circle graph below shows all Bob's expenses for last month. Bob spent $896 on his Rent last month.

5) How much did Bob's total expenses last month? _____

6) How much did Bob spend for foods last month? _____

7) How much did Bob spend for his bills last month? _____

8) How much did Bob spend on his car last month? _____

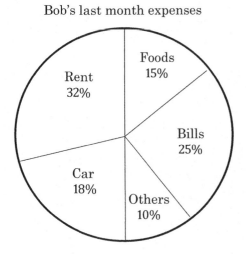

Bob's last month expenses

✎ **Solve.**

9) Bag A contains 6 red marbles and 9 green marbles. Bag B contains 4 black marbles and 7 orange marbles. What is the probability of selecting a green marble at random from bag A? What is the probability of selecting a black marble at random from Bag B?

_____   _____

✎ **Solve.**

10) Susan is baking cookies. She uses sugar, flour, butter, and eggs. How many different orders of ingredients can she try? _____

11) Jason is planning for his vacation. He wants to go to museum, go to the beach, and play volleyball. How many different ways of ordering are there for him? _____

12) In how many ways can a team of 8 basketball players choose a captain and co-captain? _____

13) How many ways can you give 6 balls to your 8 friends? _____

14) A professor is going to arrange her 6 students in a straight line. In how many ways can she do this? _____

15) In how many ways can a teacher chooses 5 out of 13 students? _____

CHAPTER 12: ANSWERS

1) Mode: 7, Range: 8, Mean: 6, Median: 7

2) Mode: 8, Range:7, Mean: 6, Median: 6.5

3) Mode: 6, Range: 13, Mean: 8.5, Median: 8

4) Mode: 1, Range: 11, Mean: 19.5, Median: 6

5) $2,800

6) $420

7) $700

8) $504

9) $\frac{3}{5}, \frac{4}{11}$

10) 24

11) 6

12) 56 (it's a permutation problem)

13) 28 (it's a combination problem)

14) 720

15) 1,287 (it's a combination problem)

"Effortless Math Education" Publications

Effortless Math authors' team strives to prepare and publish the best quality PRE-ALGEBRA learning resources to make learning Math easier for all. We hope that our publications help you learn Math in an effective way and prepare for the PRE-ALGEBRA test.

We all in Effortless Math wish you good luck and successful studies!

Effortless Math Authors

www.EffortlessMath.com

… So Much More Online!

❖ FREE Math lessons

❖ More Math learning books!

❖ Mathematics Worksheets

❖ Online Math Tutors

Need a PDF version of this book?

Visit www.EffortlessMath.com

Receive the PDF version of this book or get another FREE book!

Thank you for using our Book!

Do you LOVE this book?

Then, you can get the PDF version of this book or another book absolutely FREE!

Please email us at:

info@EffortlessMath.com

for details.

Made in the USA
Columbia, SC
25 September 2022

67920166R00072